P9-CCV-179

Run, Spot, Run

Run, Spot, Run

THE ETHICS *of* KEEPING PETS

JESSICA PIERCE

The University of Chicago Press
Chicago and London

Jessica Pierce is a bioethicist, the author of *The Last Walk*, and coauthor of *Wild Justice*.

The University of Chicago Press, Chicago 60637
The University of Chicago Press, Ltd., London

Printed in the United States of America

25 24 23 22 21 20 19 18 17 16 1 2 3 4 5

ISBN-13: 978-0-226-20989-0 (cloth)
ISBN-13: 978-0-226-20992-0 (e-book)
DOI: 10.7208/chicago/9780226209920.001.0001

Library of Congress Cataloging-in-Publication Data
Names: Pierce, Jessica, 1965– author
Title: Run, Spot, run : the ethics of keeping pets / Jessica Pierce.
Description: Chicago ; London : The University of Chicago Press, 2016. | Includes bibliographical references and index.
Identifiers: LCCN 2015038627 | ISBN 9780226209890 (cloth : alk. paper) | ISBN 9780226209920 (e-book)
Subjects: LCSH: Pets—Moral and ethical aspects. | Animal welfare—Moral and ethical aspects. | Human-animal relationships—Moral and ethical aspects.
Classification: LCC SF411.5 .P54 2016 | DDC 636.08/32—dc23
LC record available at http://lccn.loc.gov/2015038627

♾ This paper meets the requirements of ANSI/NISO Z39.48-1992 (Permanence of Paper).

Who Is Spot?

In case you're too young to know or too old to remember, Spot is a little black-and-white puppy who starred in the Dick and Jane early reader books. The Dick and Jane books were popular teaching tools during the middle decades of the twentieth century in North America, and millions of children grew up with Dick, Jane, Spot, and an orange tabby cat named Puff. The books' popularity began to wane in the 1970s, just about the same time that pet keeping began really taking off. Spot is a stand-in for all animals riding the pet-keeping wave.

Contents

A Note on Language

I am going to use the accepted language of pet keeping throughout, referring to humans as owners and animals as pets. At the end of the book, I will suggest why this language is problematic and what more appropriate language might look like. One linguistic convention I have chosen to disregard throughout is the use of the impersonal pronoun "it" in reference to specific animals.

Pet: /ˈpɛt/ (back-formation from Middle English *pety* small; first known use 1508)

1. noun. an animal kept in the domestic setting whose function is companionship, entertainment, or personal curiosity; a thing that one feels affection for.

2. verb. to stroke or pat affectionately.

3. adjective. something kept or treated as a pet.

Thinking about Spot

1. Awash with Pets

"90% of pet owners consider their furry friend a part of the family."

"Two-thirds of American pet owners are bringing their pets to bed."

"Feeling depressed? Research shows that pets are better than Prozac."

"One out of ten pets has a Facebook page."

"Owner spends hundreds of dollars on surgery for George the Goldfish."

From the looks of it, a change has been occurring in American society, and it has to do with our pets. A tidal wave of animals is surging into our homes, streets, and stores. Since about the mid-1970s, the population of pets has grown more rapidly than the human population, and the number of pets living in the United States now exceeds the number of people by a good stretch. (There are about 470 million pets and only 316 million people.) The pet wave is not limited to the United States or even to industrialized nations. Driven by greater levels of disposable income, urbanization, and evolving attitudes toward animals and toward pet ownership, the global ranks of pet animals are also swelling.

And we don't just buy pets as never before. We also treat them differently. More and more animals are living inside; it is more common to hear animals spoken of as family; we have ways of indulging our pets that would have been unimaginable forty years ago: day spas, puff coats, organic foods of higher quality than what's in our own pantries; funeral parlors; hospice care; wheelchairs and prosthetic limbs for the disabled; antidepressants and behavioral wellness counselors; stem-cell treatments and chemotherapy. Veterinarians and psychologists call it the evolution of the "human-animal bond"; social critics call it the "humanization of pets."

It seems that the fortunes of pets are at an all-time high, even while things are looking rather bleak for other groups of animals around the globe. This is perhaps why it is very rare to see the welfare of pet animals raised as an ethical concern. Pets are pampered. They are snug in their little soft beds, while lab animals and food animals and zoo animals might be chained, caged, isolated, left out in the cold or heat, and allowed to suffer all manner of other miseries. We love our pets, so why worry about them? Well, maybe love is not enough. Maybe the 470 million–odd animals we call pets also need some moral attention. Their plight may be just as serious—and perhaps in some ways even more troubled—than the billions of animals caught in the wheels of agribusiness or the biomedical research industry.

Pet keeping has dark undercurrents: the breeding facilities, the wholesale marketplaces where animals are sold like guns or toys, the high mortality, the shelters overflowing with bodies, the shockingly high numbers of animals being sexually exploited or physically abused by their owners, the punitive training methods that leave animals emotionally traumatized, the failure of more than a quarter of all pet owners to provide their animal access to basic veterinary care. Notwithstanding claims about humanization and bonding, large numbers of pets aren't getting any love or are getting the wrong kind. These undercurrents challenge even the most thoughtful and responsible pet owner because it is often hard to know what we might be doing wrong or how our actions might or might not be harming the animals we cannot see beyond the curtains of our own windows.

While many may view the increasing popularity of pet keeping as a sign that we love animals more and more, it should give us pause. Pet keeping is a tidal wave we are being carried upon—we, along with millions and millions of animals—and this wave has huge destructive potential.

2. Neighborhood Menagerie

My ideas about pet keeping have gone through a long, slow evolution. I've been thinking about and living with pets since I was a child. But I didn't really start worrying about pets until I had a child of my own who, in line with family tradition, showed a marked interest in animals and began asking for a pet This and a pet That. I was too indulgent. By the time my daughter was in elementary school, our house had been dubbed the neighborhood zoo. All the kids wanted to play at our house because there were so many animals. We had, in descending order of size (and thankfully not all at exactly the same time): dogs, a cat, guinea pigs, rats, a hamster, a snake, a salamander, a leopard gecko, a tarantula, mice, frogs, goldfish, hermit crabs, guppies, ghost crabs, miniature frogs, worms, and crickets. We also had a tadpole from a grow-your-own-frog mail-order kit (who never made it past tadpole stage), a collection of butterflies from a hatch-your-own caterpillar kit, an ant farm with live ants who were shipped to us in a little plastic tube, sea monkeys, and a kit with "Triassic" *Triops* eggs that grew into creepy-looking things that cannibalized each other and gave my daughter nightmares.

Our house was the first stop for parents whose children had grown bored with their pets. It was in this manner that we acquired guinea pigs, guppies, and several of our rats. This is also how we wound up with a cage full of mice. Mice, in my experience, do not make good pets for young children. They are too skittish, too small, and too fast. And they are excellent jumpers. A little fellow named White Yarn hopped out of my daughter's hands one day and zoomed off to parts unseen in our house. He then systematically, by cover of night, chewed away a small square of fabric right in the center of each cushion of our new couch. This act of vandalism earned him a reassignment from "pet" to "vermin."

As you can imagine, my life was absorbed with the care and feeding of so many creatures. Perhaps it was simply the exhaustion of cleaning so many cages, filling so many water bottles, and making so many trips to the pet store for rat food or aspen shavings or Nature's Miracle or live crickets. Perhaps it was the research in ethology and animal behavior I was doing during working hours that opened my eyes to the incredible richness of animal minds. Whatever the cause, I began to feel increasingly uneasy about the well-being of our captive little friends. Reading a study about goldfish intelligence made me realize how bored these creatures must be swimming in endless circles in a small bowl on the dresser. A research study on hermit crabs, which shows that they feel and remember pain, made me look at the alien creatures with more sympathy. (I cringe, thinking of the time my daughter and some friends gave the hermit crab Spidey a warm bath, which crabs are supposed to enjoy. But the water was a bit too hot and the poor thing perished.)

It could also have been the experience of getting to know Hideous Henry. One day, while my daughter and I were in PetSmart buying crickets, a store manager with whom we had become friendly asked us if we could possibly take in one more rat. Henry had been attacked by the other rats in their pet-store cage, and his whole body was covered with bites and scabs. As a hairless dumbo rat, Henry was ugly to begin with; with festering wounds, he looked truly hideous and he couldn't possibly be put out on the floor with the other for-sale rats. Henry would need to be "put down" by pet-store staff. But the manager had a soft spot for Henry and begged us to adopt him. It was during this negotiation that I learned about the store's policy that any animal who escaped from his or her cage in the store would be killed rather than sold (and such breakouts were quite common), because the store couldn't guarantee the animal to be disease free. Needless to say, Henry joined our household.

It could also have been the accumulating death toll. The garter snake only lived for a few months. The salamander survived a single summer in our care. There was Spidey the crab. There were so many small deaths, even though we had read the how-to-care-for books and were trying our best. It could have been my sense, from watching Lizzy the gecko, that trapped within the walls of a twenty-gallon glass tank was a marvelously adapted creature whose life was essentially barren, thanks to us. I couldn't help feeling sorry for her. And I couldn't help feeling sorry for the crickets she ate, who came packaged (twenty-five "pinheads" by count) in a tiny cardboard container smaller than a box of paperclips. How did they survive in there? It might also have been the day I was in PetSmart buying crickets for Lizzy and witnessed the manager receiving a Tupperware full of baby rats from a delivery man. Like my own personal road to Damascus, I was able to see. These were babies, taken from mothers.

Perhaps it was as simple as holding so many diverse creatures captive, in my own house. The feeling that so many eyes were watching me from behind the bars of the cages and the glass of the terrariums and tanks. Mini-epiphanies and small feelings of guilt began to accumulate, and I found myself becoming increasingly uncomfortable with the whole pet-keeping enterprise.

Two and a half years ago we packed up and left the old neighborhood. Several months before we moved, the last of the small critters, a sweet old rat named GooGoo, passed away. I've since been able to stand firm on a no-replacement policy. I finally developed a backbone and say no to parents and friends who try to talk us into adopting their castoffs—though I say no with regret, because it isn't the animals' fault that they have become burdensome, and I know their fate is uncertain. But I just can't do it anymore. Don't get me wrong: I may be a reformed pet addict, but I'm an addict nonetheless. We still have animals living in our house, and I can't imagine it any other way. But we have finally whittled the census down to a cat, two dogs, and two goldfish.

Klondike and Dibs live in my daughter's room, in a large tank. (They keep growing, so we keep getting larger tanks; given more space, they continue to grow. Where does the cycle end? And can they really live for twenty-five years? Will I be caring for the fish long after my daughter has left for college and started her own life?) Thor the cat joined us about three years ago, after some time on the streets and a stint at the Longmont Humane Society. He has convinced me that I am a cat person *and* a dog person. Our dog Maya is a pointer mix, now in her twelfth year, and one of the gentlest souls I've ever met. She has lived with us since puppyhood and has seen the menagerie swell and then shrink again. Bella, a tricolored mystery mutt (but most definitely with a predominance of Border collie genes), has been with us about two years and is our most recent addition. She was picked up on the street by animal control when she was about a year old, terrified and injured, and taken to the shelter. She was the first dog we saw on the day we went in with the intention of adopting a housemate for Maya. Bella has issues. (She's looking at me right now, from her head-down-butt-in-the-air position on her office dog bed saying, "Tell me about the time when you first found me and sat down next to my cage at the shelter and I growled at you. Hehehe.") She keeps things interesting.

3. Who Are Pets?

The most common pets (e.g., dogs, cats, rabbits, guinea pigs, hamsters) are drawn from the pool of domesticated animals. These species have been subjected to selective evolutionary pressures, including goal-oriented breeding by humans, over hundreds or thousands of years and have undergone significant genetic, morphological, and behavioral changes from their wild ancestors. Domesticated species take on distinctive coloration (e.g., black-and-white coats), have smaller teeth and smaller brains, and have foreshortened snouts. In some species, tails become curly and ears droopy.[1] Even more important, these animals are tolerant of human company. "Domesticated" and "tame" are not synonymous. An individual wild animal such as a lion can be tamed and socialized, if taken as a cub and raised by humans. But this lion has not been domesticated.

Domestication is often described as a process of enslavement, with humans as lord and master over animals we have tamed and made subservient to our will. Yet the domestication process is likely more nuanced: animals are not always passive victims of human manipulation and, in some cases, may have been active participants in shaping their evolutionary trajectory.[2] For at least some companion animal species, particularly dogs and cats, the domestication process has been one of mutual habituation between animal and human. And *we* have been domesticated by them, too. The particulars of how each individual species of domesticated animal came into close relationship with humankind vary, and in many cases archaeological and genetic evidence is contradictory and fails to offer a unified picture of our human-animal evolution.[3]

One recent hypothesis is that early human domestication of animals primarily focused on tameness, and that changes in physical appearance were secondary. Tameness, as an evolved trait, arises from changes in the adrenals and sympathetic nervous system, which are responsible for the fight-or-flight response. In some animals, the development of the fight-or-flight response is delayed or underfunctioning. These animals have a longer window of time—the so-called socialization window—during which humans could approach and interact, without the fight-or-flight response being fully activated. By the time the adrenal glands and sympathetic nervous system develop, the animal is already habituated to humans and is, accordingly, quite tame. The neural crest cells that form the adrenal glands and parts of the nervous system also determine pigmentation and the development of skull, teeth, and ears. If these fight-or-flight challenged animals were selected for over and over, the species would become increasingly tame and would also show distinctive changes in appearance.[4]

This is only one possibility, of course, and research will continue to shed light on exactly what sorts of genetic changes have been taking place during domestication and what their reverberations are for animals. Scientific debates about how and why domestication occurred may seem purely academic, but they have broader relevance. They inform our understanding of who our companion animals are and often have practical fallout for pets and pet owners (e.g., when dog-training methods are based on the behavior of wolves).

Although domesticated animals, and particularly dogs and cats, are our most familiar pets, human pet keeping is hardly limited to domesticated species. We also make pets out of wild species, some of which we breed in captivity (leopard geckos, ball pythons), and some of which we capture from their home ecosystems (tortoises, blue macaws, monkeys). We make pets out of other mammals, as well as out of reptiles, amphibians, birds, fish, and even insects.

The only thing limiting the kind of pet humans are willing and eager to keep is our imagination, and some people seem drawn to the most bizarre animals they can find. Once you imagine what you want, you will have little trouble finding someone willing to sell it to you.

Although we make pets out of nearly anything that moves, and even things that don't, it might be said that some creatures make better pets than others. This, of course, depends on our definition of "pet." If we view a pet as an object or source of interest and entertainment, and if the relationship is purely one directional, then the more different the better, perhaps. But if we view pet keeping as the formation of a meaningful friendship or social bond with an animal—and if the animal's perspective on the whole affair is important to us—then domesticated species with the most behavioral and cognitive similarity to humans, and species like dogs that have coevolved *with* humans, may make the best pets.

4. Why Pets?

I ask sometimes why these small animals
With bitter eyes, why we should care for them

Jon Silken, "Caring for Animals"

How humans form social attachments to nonhuman animals is well understood. But why we should be so drawn to keeping animals as pets remains elusive, as does that mysterious je ne sais quoi that makes us declare a particular animal "pet" as opposed to dinner. Nearly all infants and young children have an innate curiosity about and interest in animals, and there is no question that animals hold a magnetic attraction for the very young.[1] James Serpell, one of our foremost experts on human patterns of pet keeping, argues that pet keeping has a long historical pedigree and, while perhaps not found in every single culture over time and across the globe, comes pretty close to being a human universal.[2]

But there is huge diversity, both within cultures and between them, in how older children and adult humans relate to animals and, particularly, to pets. Some love animals, some feel fear or disgust; some love dogs and hate cats; some love snakes and lizards, while others find them creepy. Some of this diversity in animal feeling is surely related to enculturation (we in the United States tend to love dogs and be disgusted by the thought of eating them; other cultures farm dogs like we do pigs and relish the thought of grilled dog with hot chili sauce). And some is related to personal experience (dog lovers are statistically more likely to have grown up in homes with dogs than not) and serendipity. So we wind up with a notion of pet that is quite jumpy and slippery, like a little mouse in a child's cupped hands. As soon as one definition is offered, counterexamples spring to mind. The most we can say is that "pet" is an arbitrary assignation, a social construct.

The dictionary definition of "pet" is a domestic or tamed animal kept for pleasure or companionship and treated with affection. Following this definition, it has been generally claimed that pet is a category of animal with no economic or utilitarian function, setting the pet into a wholly separate moral category from animals raised for food, laboratory research, hard labor, or even to be put on display in a zoo. Our utilitarian animals are viewed and treated as things, as units of production, which perhaps helps to explain the ethical lapses in our care of them. Pets, in contrast, are doted on and treasured. Our relationships with them are affiliative, and they are like family. Rather than mere things, they enjoy a moral status verging on "quasi personhood."

Serpell takes this line of argument. Pets, he says, serve no useful function and are thus exempt from our utilitarian calculations. "Fondness for pets in all societies is largely independent of the animal's contribution to the local or family economy."[3] To illustrate, he compares pets with food animals. After a long description of the horrid and brief life of a pig destined for the bacon aisle, he writes, "This kind of hard-nosed, economic attitude to the exploitation of domestic animals is a simple and straightforward one, and it is one that is tacitly endorsed by the majority of people in the western world. Humans have a right to eat meat; farmers have a duty to supply this demand as cheaply as possible; animals inevitably suffer as a consequence." Yet, he goes on, "there exists in our society an entirely separate category of domestic animals which, for no obvious reason, is exempt from this sort of treatment."[4] Our pets. But pet animals are not exempt from exploitation, and they fall victim to the exact same kind of economic calculus as pigs destined for the grocery store. Pets are not eaten (usually), but they feed our souls. The right to keep animals as pets is tacitly endorsed by the majority of people, and suppliers duly line up to fulfill this demand. And animals inevitably suffer as a consequence.

For the individual pet owner, the benefits of pet ownership rarely boil down to the economic. My pets bring me no financial gain whatsoever and, in fact, impose quite a burden. But to remain focused on the individual's reasons for pet keeping is to ignore the bigger picture: pets are an enormous economic benefit to someone, indeed to lots of people. And the economic forces that encourage our pet-keeping habits and that make animals widely, easily, and cheaply available are surely one of the key driving forces behind the tidal wave of pet keeping. The pet industry helps to create the social conditions within which pet keeping is a highly valued activity.

Yet still, beneath all the marketing and social construction of attitudes and desires and shopping habits, there is a deep yearning for connection. The main reason people seek the company of pets must be psychological: animals make people happy and satisfy a basic urge to tend, to love, to bond. "Companion animals," write Henri Julius and colleagues, "may satisfy the need of individual humans for a reasonably compassionate partner . . . whom they can care for and attach to, at comparatively low 'social costs.' For example, cats and dogs do not argue verbally and are less demanding in many respects than a human partner." Relationships with animals emphasize the emotional component, downplaying cognitive and cultural components that can complicate relationships between humans. Companion animals adjust "asymmetrically and uncompromisingly" to their human partners. "The choice of an animal [for a pet] may depend on its quality as a target for projecting attitudes and desires."[5] Socially intelligent species are "open program," so social behavior is based on learning and experience. They can be enculturated to fit a human social environment and to respond to us in the ways we desire. We consider it ideal to acquire a companion animal at a very young age so that he or she can be properly "socialized." The socialization process involves getting them to imprint on us, rather than on their own parents; we want them to bond with us, and just with us.

So we come around, again, to an uncomfortable sense that we may *use* pets, even if emotional fulfillment is a relatively benign form of exploitation. I recently asked a veterinarian friend how he thought about our relationship to companion animals. "They are slaves—chattel slaves," he said breezily. I was astonished and irked by the nonchalance of his proclamation. But as I thought about it afterward, the reason his statement bothered me so much was that he had touched a nerve—had captured something I find uncomfortable about our relationship to pets. They are here to serve us.

Geographer Yi-Fu Tuan explored the ambiguous dynamic of pet keeping in his 1984 book *Dominance and Affection*. Keeping a creature as a pet is an act of domination, he says, but because it takes places within the realm of play and pleasure and because it is cloaked in affection, this form of pathology has flown under the radar of our attention. Volumes have been written on the abuses of power in the economic and political realms; as for abuses of power in the realm of play, there is almost complete silence.

> Dominance may be cruel and exploitative, with no hint of affection in it. What it produces is the victim. On the other hand, dominance may be combined with affection, and what it produces is the pet.[6]

The psychology of playful domination is that "warm and superior feeling one has toward things that one can care for and patronize."[7] Much as I would like to deny it, his sentence has a ring of truth. Sometimes when I look at Thor the cat I think, "Yes, I have made him a slave to my own whims and desires, particularly my desire to enjoy the company of an intelligent and emotionally sensitive warm-blooded creature." Before he became "mine," Thor was the property of the Humane Society. I adopted him (technically, I bought him for $80), and now he belongs to me. And I do, indeed, get a warm feeling from having a friend in the house, and from tending to Thor's care. Although I think Thor is happy in our household, he really didn't have any choice but to join us.

Our relationship with our pets may involve domination, but this is clearly not the full story. Anyone who has experienced a close bond with an animal knows this deeper truth: these are not mere objects, and affection is far too facile a description of the feelings we have for them. My feelings for Thor, and I think his for me, go well beyond affection. Sometimes a human-animal relationship approaches a point of symmetry and—dare we say it—of equality, where human and animal share, by mutual choice, the same physical and emotional space. Although Thor's reign is limited to our home and our hillside, within this space he shares in the social life of the family on his own terms. Because the door is open for him, he is free to go, and yet he chooses to stay.

5. Tainted Love

It is easy to stay focused on the love relationship and to bask in the joy of our own pets. Yet even for those who consider animals part of the family, living with them is not always sunny, despite what television commercials try to tell us. There can be stress, heartache, failed relationships, a sense of misgiving, a feeling of wanting to do best for our animals yet not being quite sure what "best" is. Our love affair with animals is complicated.

In this introductory section, "Thinking about Spot," I've tried to show why pet keeping is an ethically rich and important area of investigation. If an issue is cut-and-dried, it doesn't make for compelling book material. You state your opinion and be done with it. Far more interesting are the issues that, once you begin thinking in more detail, blossom into something multifaceted and complex, with no clear right or wrong answers. This is what the ethics of pet keeping is like.

I would guess that there are things about keeping pets that make you uncomfortable. Maybe you've had to decide how much to spend on veterinary care (what is my animal's life worth, as a percentage of my paycheck?); or maybe you worry about the suffering of food animals, yet believe that a carnivorous diet is the only healthy option for your cat; or perhaps you have to leave your dog alone everyday while you go to work, and you are concerned about him being lonely and getting enough social stimulation and physical exercise. What do our animals need from us, in order to be happy? Can you have a fondness for pets yet question the morality of holding them captive? The next section, "Living with Spot," explores these and other questions of individual responsibility, focusing on the private sphere of the home and neighborhood.

The third section, "Worrying about Spot," reaches beyond the individual pet owner to consider the larger impacts of our pet-keeping obsession on the animals caught up in this tide. Many aspects of pet keeping—captivity, confinement, boredom, abuse, abandonment—are very hard on animals. Furthermore, the cultural practice of pet keeping has reverberations well beyond the individual pet owner and his or her animals. Like many other social consciousness-raising projects, I hope to shed some light on the global implications of our local practices. By "global" I mean for pet animals writ large; my focus will remain, with a few exceptions, on pet-keeping practices in the United States. What happens when animals become commodities? Where do the animals on the pet-store shelves come from? What happens to the surplus animals and to the ones people buy but no longer want?

In the concluding section, "Taking Care of Spot," I'll ask whether pet keeping can withstand the rigors of moral examination and will explore what more ethical pet-keeping practices might look like, for us and for animals.

I would like to suggest that companion animals are an important focus of ethical concern and here's why:

- there are millions of pets and their numbers are increasing;
- the majority of people think that pets are fine and that there is nothing they need protection from, and all the while real welfare concerns slip by under the radar of consciousness;
- exploitation and cruelty are rampant within the pet industry and among many segments of the pet-owning public, and animals are suffering;
- pathologies of violence toward animals are intimately tied to violence against people;
- and finally, harm to animals can be addressed on multiple fronts and we can hope for some synergistic effects.

Awareness of the plight of pets, as animals with whom we most clearly empathize, may increase concern for all animals. And who can argue against a world full of greater compassion for all creatures?

Above all else, I hope that these chapters and the whole they create will offer you a chance to consider the animals' perspective on pet keeping. Of course, we cannot adopt "the animals' perspective." Animals are not a singular entity; even "cats" are not a homogeneous unit. Perhaps, then, we can pursue a more modest goal: let's acknowledge that each animal we acquire and keep as a pet is an individual whose own perspective on being a pet we ought to consider. We cannot become another animal, but we can try walking in their paws and can take an imaginative journey into their world. I'm afraid we won't like what we see. Once we've seen, then we need to begin speaking out, because silence is a form of acquiescence. Our animal companions need our voice of protest and protection.

Living with Spot

6. Family Constellations

"Pets are part of the family!" is a mantra of the media, the pet industry, and even the academic veterinary literature. And, indeed, a good number of pets *are* considered part of a family. Pets are often part of what some psychologists call the family constellation, an image that I like because it suggests a dynamic system held together or pulled apart by unseen but powerful gravitational forces. In a well-known study by Sandra Barker and Randolph Barker from the 1980s, dog owners were asked to complete what is called the Family Life Space Diagram, in which symbols representing family members and dogs are placed within a drawn circle representing one's "life space." In 38 percent of the diagrams, the dog was placed closer to the self than were other family members.[1] Similar studies of pets' placement within a family constellation have similar results: pets are quite often drawn very close to the center—closer even than human family members. Asked which one person they would bring to a desert island, a surprising number of people say "my dog" or "my cat" rather than "my husband" or "my wife." This doesn't mean we love our pets more than our spouses or children, or that we consider our pets to be people. What it means is that our pets often share our life space more closely and fully than other humans. This is true for me and my critters. I couldn't visualize my life space without including Bella, Maya, and Thor. Because I work at home, I spend far more time with my dogs than I do with my husband and daughter, who are off at the office or school for many hours of the day. My human family members are independent of me—they have their own life spaces—while also overlapping with mine. My animals are not very independent, so their life spaces fit snugly within my own (or at least this is how I imagine it).

An animal's status is oftentimes shifting, fleeting, changing. Ethnographic research shows us just how tenuous human-animal bonds can be. Maybe the relationship becomes strained by what the human perceives as "behavioral problems" in the animal, or maybe there are changes in the human's situation (divorce, illness, loss of job, new baby) that make the animal's presence inconvenient. Either way, the animal is often simply ejected from the family system. As Israeli anthropologist Dafna Shir-Vertesh phrases it, animals are "flexible persons" or "emotional commodities." They are persons when we want them to be, and when we tire of them or they create tension in a family, they are demoted to "just a dog."[2]

Pets are often spoken of as children. Some people call their pets "furbabies," and more often than not, the pet industry, and even now an increasing number of veterinary practices, are referring to owners as "pet parents." Like small children, pets must be continually cared for. They must be fed, watered, kept from self-injury, groomed, taken to the doctor, and generally looked after. Dogs, especially, have been genetically shaped to look and behave like juveniles and to activate our caregiving instincts.[3] And unlike our children, who at some point begin breaking away, even refusing our caregiving advances, we are assured of the pet's continuing dependence. As Alan Beck and Aaron Katcher write,

> Apparently, in our best and most innocent affectionate interchanges with our pets, and in some of the worst, the animal still acts as a kind of child . . . [and] defining a pet as a kind of child permits us to nurture the animal and gain all of the benefits that such nurturing provides. . . . The little acts of caring, feeding, watering, tending, and protecting all call forth a response, and the sum of the

> acts leaves the caregiver with the feeling that she is needed. The reciprocal feelings of caring for something and being needed are lines that can hold us to life.[4]

Michael Robin and Robert ten Bensel have suggested that pets, when first acquired, function like new family members.[5] With the arrival of a new pet, families undergo various changes, some good and some bad. Getting a pet may bring a family closer, by encouraging time together, or it may exacerbate dysfunctions, for example, by generating disagreements over responsibility for care. Difficult family dynamics can be made worse by the presence of a pet, for instance, by a dog who won't let a husband near his wife (and maybe she likes it that way). Animals can become rivals. In a letter to "Dear Abby," to give an example, a woman bemoans the fact that her fiancé constantly says that he is too tired or busy to show her affection, yet he "goes out of his way to snuggle and play with our two dogs and cat."[6]

In families where domestic violence is occurring, a pet can become a child's sole object of love; the animal can and often also becomes both a tool and a victim of violence (see chap. 30, "The Links"). Pets are often introduced into a family in order to create a bridge between parent and child. Sometimes it doesn't work, though: the animal doesn't do his or her job. Sometimes the parent then gets rid of the pet, or a child bonds only with the animal and shuts out the parent. Unwittingly, pets become part of a family's dysfunction.

Within a given family, there will be a whole range of relationships with the pet. So, to label a pet a full-fledged member of the family is far too simplistic. It may be that the mother has a deep and codependent, even pathological, relationship with the animal; the husband actively dislikes the animal; the daughter loves the animal and feels a close bond and also uses the animal's clear preference for her over her brother as a form of sibling competition; the son may view the animal as a plaything.

Pets also become involved in family "triangulation." In family systems theory, a two-person family system is inherently unstable and will, when placed under a certain amount of tension, form a three-"person" triangle.[7] Pets can be triangulated into a family system to relieve tension or to create conflict. An example of triangulation involving pets might be someone taking anger or tension out on the pet instead of their spouse. Someone might turn to a pet for comfort rather than their spouse, and in this way fuel resentment against the animal. A pet can wind up in the middle of human fighting (for instance, the husband sweet talks the cat to distance his wife; husband and wife focus their affection on the cat rather than on each other; the wife talks through the cat as "mouthpiece" instead directly to her husband: "Fluffy, can you believe that he forgot to take out the trash? Again? I guess you and I are going to have to do all the work ourselves, aren't we?")

Our pets may be part of the family, for better or worse, but to what extent is our life sharing reciprocal? One thing you hear frequently, in relation to the pets-as-family narrative, is that humans and their pets are so close because we share in many daily activities. This is particularly true for those of us who live with dogs because they can participate in a number of activities outside the home sphere. They can go walking or jogging with us, can go in the car to do errands, or join us on vacations. But in what sense are these activities really shared? Although a person and a dog may share in the performance of an activity (walking, agility trials), in nearly all cases, the instigation and termination of the activity are dictated by the human.[8] During a given day, Bella makes hundreds of play invitations to me by dropping a ball in my lap while I'm sitting at the computer. I accept a few of these, but not many. And actually, I will most often ignore her requests and instigate my own invitation when I am good and ready. My thinking goes like this: if I accept her invitations, I'll reinforce a potentially annoying behavior. And this is, indeed, what behavior books will recommend: I must teach Bella that I will be The Instigator. Yet if we are always the boss and always calling the shots, just how mutual and reciprocal are our relations with animals?

Do animals create family for people who are otherwise socially isolated? Sometimes, yes. An animal can be a lifesaver for a person who is lonely, whether because they have lost a spouse, or their children have grown up and moved away, or simply because they don't connect well with members of their own species.[9] But people living alone actually make up the smallest pet-owning demographic. The largest demographic is families, particularly those with young children.

My suspicion, without seeing research specifically directed at this question, is that cultural pressures to "complete the family" are strong, and the idea that having a pet is a good thing for children drives a number of these pet acquisitions. So, too, does the practice of acquiring a dog or a cat as a kind of "starter child"—a first born, to create a nuclear family and practice your child-rearing skills and work through potential parenting differences. When the children come along, the dog or cat may lose his place, both emotionally and often quite literally. Ironically, it is precisely the young family who should be most cautious and most resistant to getting a pet. They are likely to have less time, less money, and fewer reserves of patience (said reserves being used up quickly by the human offspring). The older the children get, the busier life will be and the more challenging to provide appropriately for an animal. And indeed, in studies of attachment to pets, young families report the lowest levels of attachment and are often the ones who wind up relinquishing their animal to a shelter.

One final thing to say about interspecies family constellations is that bringing a new animal into a home that already includes one or more pets can have profound and unanticipated reverberations throughout the system. For example, going from a one-dog household to a two-dog household is a significant shift, adding multiple layers of complexity to the constellation (the complicated relationships between the dogs, the vying for attention, the unique connections between each dog and each person). Although "sibling" dogs will often develop a deep bond of friendship, sometimes they never adapt to each other, and a significant number of visits to veterinary behaviorists are for intercanine aggression within a household. In many cases, the fighting is vicious and intractable, and one or more of the dogs wind up being rehomed or euthanized. The same is true for cats, who may or may not accept a new feline family member. A quick meet and greet before adoption is advisable—and many shelters require this—but initial signs of acceptance offer no guarantee of harmonious future relations. Adding new species, such as introducing a cat into a human-canine household, requires a whole additional level of reorganization.

Although many pets are included, for better or worse, in family constellations, a good many are not, and the pets-are-family claim is far from a universal. In fact, the "data" on high levels of inclusion of pets in families come from questionnaires that sample only a certain pet-owning demographic, and the results are quite skewed. Nothing from these surveys can be generalized to the entire population of animals living as pets in our society. Animals used for breeding stock are not part of a family, nor are the homeless animals in our shelters. Emotional attachments to and functions of animals in human families are highly variable—and not always positive. Some family pets are surrogate children, some are partners, some are tangible property, some are unpaid employees, some are scapegoats, and some are targets for emotional and physical or sexual violence.

7. Why Not

Two stereotypes bedevil our pet-obsessed culture: pet owners love animals and want to care for them; non–pet owners dislike animals, are selfish, and don't want to be bothered. The first stereotype is challenged throughout this book. So let's take a look at the second.

Yes, many non–pet owners dislike animals. I'm not sure quite how it happens, but some people actually think animals are dirty and disease ridden and just plain irritating. If you have pets, you've certainly encountered your share of Haters. But there is a whole other cohort of non–pet owners who love animals. Furthermore, some of the most responsible pet owners I know don't have pets. Let me explain by way of our friends Julie and Mark. They both love animals. They had a huge, slobbery Saint Bernard named Ben, who was part of their family for ten years and who they treated like a king. Now that Ben has passed away, their three children beg and beg for a new puppy. Julie says no and she is adamant. "We just don't have the kind of life right now that can accommodate a dog. We are too busy. The kids all have their after-school stuff, we're both working, we like to take trips . . . it wouldn't be fair to the animal."

Research exploring why people keep pets and why they don't offers some surprises. As you'll see, Julie and Mark's reasons are shared by others who have made a conscientious choice not to have a pet. In *The Sciences of Animal Welfare*, David Mellor, Emily

Patterson-Kane, and Kevin J. Stafford compiled results from several studies to offer at least a partial answer. They provide the following summary[1]:

Reasons people give for owning a pet
(in descending order of importance)

Companionship
Love and affection
For the children
Someone to greet me
Property protection
Someone to care for
Beauty of animal
Sport (e.g., hunting)
Show value

Reasons people give for *not* owning a pet
(in descending order of importance)

A problem when I go away
Not enough time
Poor housing for pet
Location dangerous for pet
Pets not allowed
Family allergy problems
Dislike animals
Too expensive
Zoonoses [i.e., diseases that can be transmitted to humans from animals]

What's interesting is that those who own a pet cite mainly owner focused or selfish reasons. Non–pet owners cite animal-related reasons, mainly concerns over adequate welfare of the animal. Many of the things appearing in the "reasons for owning a pet" category can become significant concerns for an animal. For example, the amount of time the average pet owner spends with his or her critter is estimated at about forty minutes a day—hardly enough to be called a reciprocal and mutually enhancing friendship.[2] Furthermore, acquiring an animal as a child's toy can play out in unexpected ways: the child doesn't bond with the animal; the animal creates unexpected family conflicts and increases emotional stress rather than decreasing it; the child gets older and gets involved in his own activities and doesn't have time for the animal. Likewise, animals used as emotional crutches are sometimes made so dependent on their owners that they cannot function when left on their own (thus the epidemic of anxiety disorders in our nation's dogs).

As I've noted elsewhere, demographic studies of pet ownership show that young couples and families with young children are the groups most likely to acquire a pet. They are also the groups who, according to research, are least well bonded with their animal and have the least strong commitment. Driving this trend may be the cultural perception, thoughtfully molded by pet-industry marketing, that a pet is a necessary part of a happy family. Since we live in a culture that likes to medicalize things, perhaps we need a diagnosis: "pet-deficit disorder." You can find relief at your local pet store.

8. Sleeping Together

It is most certain that the breath and savour of cats consume the radical humour and destroy the lungs, and they who keep cats with them in their beds have the air corrupted and fall into hectics and consumptions. . . . The hair of the cat being eaten unawares stops the artery and causes suffocation.

Edward Topsell, *The History of Four-Footed Beasts* (1607)[1]

I have slept with my cats for two decades—and would much rather go to bed with them than with a man.

Anonymous

Roughly half of all pet owners share a bed with their pet, according to a WebMD survey. That's about all we know regarding the practice of interspecies bed sharing. But where research is lacking, strong opinions are not. Some decry bed sharing as dangerous and disgusting; others consider it one of the finer pleasures in life. I tend toward the latter. I love sleeping with my dogs and cat, though it often means I don't get good rest. There is the excitement over a noise heard outside the window, with barking and jumping on and off the bed. It gets hot having so many warm-blooded creatures clumped together. And then there is Thor, whose internal kitty alarm clock is set for 4 A.M., at which time he walks on top of our heads, back and forth a few times until one of us finally gets up or shuts him out of the room.

There is only one argument to offer in favor of bed sharing, but it is a compelling one: it feels so right. Our critters are warm and cuddly, and it's good to be part of a pack.

On the down side, we have the potential for sleep disruption, zoonotic diseases, and bad behavioral habits.

A small study conducted by the Mayo Clinic Center for Sleep Medicine found that about 10 percent of their pet-owning patients reported "annoyance" that their pets were disturbing their sleep (up from only 1 percent ten years earlier, perhaps a sign of increasing numbers of pets in households). Disturbances included noises (whimpering, squawking, barking, growling), needing to let an animal outside, and attending to medical needs.[2] As you'll notice, these disturbances are not related to bed sharing, per se, but simply to the presence of pets in the house. Pets will likely disrupt our sleep whether or not they are allowed in our beds, and there isn't much we can do about it.

This particular burden of pet ownership rarely gets the attention it should, and I suspect that many animals wind up in the shelter because of unforeseen problems of this nature. Sleep challenges should most certainly be added to the list of things people need to think carefully about *before* bringing a pet into their home. One quite common experience: a child wheedles her parents into getting a rat or hamster or guinea pig, ostensibly so the child won't be afraid to sleep in her room alone at night. Well, guess what? Most of the aforementioned critters are nocturnal and will keep little Susie awake with their food gnawing and wheel running and litter arranging. I fell into this trap myself. This is how my daughter's rats came to inhabit the dining room. We lived in a small house, and when we had dinner guests, they would be seated right next to the rat hotel. Perhaps this is why friends often suggested eating out.

If you think about where your cat or dog has been and what she has done during a given day, concern about the spread of germs and zoonotic diseases seems quite reasonable (see chap. 15, "Cat Scratch Fever"). On one particularly memorable night, I woke up to find Bella barfing on me, and the entire bed littered with chewed up bits of a desiccated animal carcass that she had somehow sneaked into the house. An event such as this notwithstanding, our risk of getting a dread disease from our pets isn't related to bed sharing, in itself. Just having them in the house puts us at risk, especially kissing them on the nose or mouth or getting scratched or failing to wash our hands after a vigorous petting session. Bed sharing probably increases our risk, because it puts us into extended close contact. As the folk saying warns, "Go to bed with a dog, you might wake up with fleas." But many people, including me, feel the risk is worth the reward.

Because the bed is often one of the most coveted pieces of territory in a home, it can also become a source of conflict. Behaviorists warn that allowing a dog in the bed can sometimes lead to aggressive behavior, directed either at other animals or at humans.[3] In multidog households, the bed can easily become a resource to be guarded and fought over—sometimes even peed on. I haven't seen much discussion of behavioral issues raised by sharing a bed with a cat, but in multicat households, territorial issues could be exacerbated by allowing one or more of the cats to sleep in the bed. Similar (or very different) kinds of issues may arise with the various other critters who share people's beds, such as parrots, hamsters, and pot-bellied pigs.

9. Stroke Me

One of the great pleasures of having pets is petting them—running our hands through their fur, rubbing a belly, or just laying a hand on their back and feeling the warmth. We love to touch our animals and they seem, for the most part, to enjoy touching us and being touched by us. What is a pet, after all, but something to stroke? The exact origin of the noun "pet" is unknown, but the word came into use in the sixteenth century and referred to an indulged or spoiled child or to a tamed animal. The verb "to pet" was first used in the seventeenth century and means to stroke or fondle. (The erotic use of the term, as in "heavy petting," is a modern iteration.) My animals seek out my touch throughout the day, and I theirs. One of my favorite ways of interacting with Maya, Bella, and Thor is through touch. I love it when Thor curls up on my desk, right in front of or on top of my computer keyboard, so I can scratch his head and stroke his fur while I pretend to work.

Biologists have discovered physiological explanations for why we enjoy touching and being touched. Consider, for example, what are called stroke-sensitive neurons. A small population of sensory neurons called C-tactile afferents can be found in hairy-skinned mammals (and yes, that includes us). Gentle stroking of these neurons stimulates the release of the hormone oxytocin. Oxytocin plays an important role in maternal behaviors, intimacy, pair bonding, grooming behavior, trust, and wound healing. As their name suggests, these neurons respond to gentle stroking but not to pinching or poking.[1]

Gentle stroking (of the right sort—see chap. 31, "Heavy Petting") has marked physiological effects both for the stroker and the strokee. During petting, human and animal both get a surge of oxytocin. A dog being petted by her owner shows lowered levels of stress hormones and a lowered heart rate. A person touching the fur of a dog or cat likewise shows a decrease in blood pressure and a lowering of heart rate. Touching also releases endogenous opiates such as endorphin, which can decrease pain and make us feel good. Touch plays a central role in the human-animal bond.

Social animals are sensitive to physical intimacy, but what feels good and what feels invasive will vary a lot from one individual to another and will also vary from species to species. Research on cats, for example, has shown that some get stressed out by being petted.[2] Other cats seek frequent physical touch by their owners. Some dogs enjoy the touch of their own human, but are wary of strangers and don't like to be touched by them. Small animals, such as guinea pigs, sugar gliders, and rats, can be become stressed if they aren't used to being handled or if the handling is rough. Touching can get invasive: think of a time when someone got too close or touched you in a way that made you feel uncomfortable. Animals can get to feeling uncomfortable, too.

We often seem to mistake our pets for stuffed animal toys. We cuddle and squeeze and stroke, and then cuddle and squeeze and stroke some more, assuming that all the attention we shower on them is lapped up like gravy. Anyone who has raised children will understand the feeling of simply needing some personal space and the desperation that can set in when someone—even someone you love—spends too much time crawling all over you, stroking your hair, fondling your face. You find yourself explaining the concept of a space bubble, and how everybody has personal space that needs to be respected. Often we jump to the conclusion that a dog or cat who doesn't like to be petted is cranky or unfriendly, but this isn't necessarily true. They might just like more personal space. Animals need the chance to set their boundaries and say no, just as we do. This is difficult when you are on a leash, or held in someone's arms, or are in a tiny cage.

Patricia McConnell explains, in *The Other End of the Leash*, that sometimes there is a mismatch between dog and dog owner. One is aloof and the other is touchy-feely. To a human, aloofness in a dog might be read as dislike. To a cuddly dog with an aloof owner, the need for social contact may be going unmet. And even for a dog who likes to be touched, there are times when touching may not be appropriate (e.g., when the dog is in the middle of a play session or when he is highly aroused or is sick).

What if we want to touch an animal but don't have a dog or cat handy? One of the inventions that made a recent *New York Times Magazine* list of "innovations that will change your tomorrow" is called smart fur. Developed by researchers at the University of British Columbia, smart fur looks like nothing more than an odd little scrap of faux fur or maybe a misplaced toupee. Built into the fur, though, is a sensor made of conductive threads. When you touch it, it actually feels like a real animal. What is even more interesting is that the smart fur can "read" human emotion through the way it is touched and researchers say that it can differentiate nine different emotional gestures. Smart fur is part of the latest trend in "haptic creatures" (haptic refers to our sense of touch; haptic designs incorporate tactile feedback into the user interface).

Smart fur isn't just for fun. The basic idea behind smart fur is that people love to pet animals, and petting animals has therapeutic benefits. There are many instances in which petting might be beneficial but where interacting with a real dog or cat is not feasible. For example, smart fur could be used with hospitalized patients or elderly people in nursing homes. It could also be used by animal lovers with severe allergies. And it could be used by ordinary people like me. I could keep a smart fur in my purse and stroke it when stuck in traffic or in a long grocery store line or in other stressful situations. I definitely want a smart fur. Eventually, the smart fur might be combined with a "robo-pet" to create something even more animal-like. A prototype smart fur roborabbit is currently under development, paving the way perhaps for a whole new type of pet ownership, free of the welfare concerns related to real live animals.

A discussion of the role of touch in the human-animal relationship is not complete without mention of the critical role of human touch in veterinary medicine, particularly in some of the complementary healing modalities. As I noted before, touch releases endorphins and generates a cascade of beneficial effects: relaxing of the muscles, increased circulation, and increased flow of oxygen, all of which stimulate the body to heal itself. Healing Touch is a well-accepted holistic modality within human medicine and works through the body's energy centers (chakras) and through energy outside the body. Healing Touch for Animals (HTA) draws on the same therapeutic principles (though the energy fields of animals are said to be different from human energy fields, so special training for veterinary application is useful). By placing hands gently on or very near the skin of the animal and directing intention to the flow of energy, a therapist or pet owner can sometimes stimulate a healing response. Tellington Touch (TTouch), which uses gentle stroking at the acupressure points, and massage, which involves gentle manipulation of the fascia, can also be effective in the care of animals. All of these touch-based therapies can be helpful to animals who are ill, anxious, traumatized, or suffering from pain.

10. Talk Talk

Lots of people talk to animals. Not very many listen, though. That's the problem.

Benjamin Hoffman, *The Tao of Pooh*

I talk to my animals all day. I keep up a constant stream of conversation, most of it embarrassingly banal: "Let's go see if the laundry is done, shall we? Then we'll get some chocolate for me and a biscuit for you." And apparently, I'm not crazy or, if I am, at least I'm in good company. A vast majority of pet owners report talking to their animals—a lot. Do our animals listen? They don't really have a choice. They are a captive audience. But nonetheless, I do think they listen, and in ways that seem to be important to their human companions. Animals listen without judging, without kickback, without interrupting us to talk about themselves. You can pour your heart out and know that your secrets really are safe. You can reveal yourself without pretense, without caution. But do they *hear* what we're saying? Given the emotional perceptiveness of dogs and cats and other animals, they likely understand quite a lot, even though we speak in a foreign tongue.

But talking to our animals isn't really "communicating." Communication is a reciprocal activity, a passing back and forth of relevant information in a form that can be understood. One of the most important responsibilities of a pet owner is learning to communicate clearly by learning to "listen" to our animals and learning to "talk" clearly so that they can understand what we are asking of them. Being able to communicate well is the foundation for a successful relationship, and miscommunication can lead to heartache, bad feelings, and failure to bond.

We shouldn't underestimate how much can be communicated nonverbally. We are a highly verbal species, but as all of us know, words are just that; they are only the surface. It is how somebody says the words, or when, or with what look in their eye that tells us what we really need to know. In fact, scientists studying humans have estimated that between 60 and 90 percent of communication in a verbal exchange is actually nonverbal. So the notion that we can't communicate with other animals because they can't speak English or Spanish is absurd. When communication failures occur, we typically blame it on the animal. But it is more likely our own fault. Dog bites are a classic example of our failure to listen. Dogs rarely just lash out and bite; they give lots of warning communications. But people often fail to notice, especially children, who are less adept than adults at reading dogs' behavioral cues. Children are especially likely to misinterpret dog facial expressions and will, for example, interpret teeth baring in a dog as laughing. A child trying to hug a "laughing" dog may get a nasty surprise.[1]

Learning to communicate with an animal companion is a bit like learning a foreign language: there is a whole new vocabulary, and sounds and even gestures carry different meanings. Dogs, for example, communicate through a variety of barks, growls, whines, and whimpers. They talk with their ears, eyes, face, and body. (Dogs who lose their tails to docking lose one of their key tools for communicating.) Dogs and cats both also have a specialized structure in their nasal passage called the vomeronasal organ, which picks up chemical signals called pheromones. They communicate with each other this way—especially about sex—and can also "smell" human pheromones much more effectively than we ever could. Humans communicate through chemical signals as well, but our skills in this area are vastly inferior to those of most companion animal species.

If, like me, you have both dogs and cats, you have significantly more homework to do. Cat and Dog are two completely different languages, and knowing how to read a dog won't necessarily help us translate the mysteries of the feline mind. In dogs, for example, a wagging tail is (usually) a friendly invitation to come closer. A great big wag is like a great big smile. In cats, a wagging tail communicates that a cat is feeling uneasy. The broader the swing, the more aggressive and less smiley the cat is feeling. Luckily, dogs and cats who grow up together or who live together over many years will generally be quite adept at adapting to each other's communication styles.

The ability of humans and dogs to communicate with each other is nothing short of remarkable. During their evolution, dogs have acquired the skill of reading human faces and emotions.[2] They consistently show a left bias in looking at human faces. They know to follow a pointing finger, and they follow the direction of our gaze. They can even appear to make inferences about the reliability of the human giving the cues, responding more consistently to reliable human sources.[3] Researchers believe that dogs evolved their vocal repertoire to facilitate communication *with us*. And we are actually quite adept at reading their signals, too. For example, humans are good at accurately interpreting dog emotions by quality of barks.[4] And just as dogs follow our gaze, they seem to expect us to follow theirs. Bella demonstrates this many times a day. She will put her ball or Frisbee down where I can see it and then focus a laser stare at it, flicking her eyes up every now and then to see whether I'm getting the message.

When it comes to behavioral cues that are not explicitly directed at us by our animals, humans do somewhat less well. For example, a 2012 study by veterinarian Chiara Mariti and colleagues found lapses in how well owners perceive stress in their dogs. While owners were good at identifying trembling and whining as signs of stress, they rarely reported more subtle behaviors such as averting the head, yawning, and nose licking.[5] These lapses are important because they directly affect the well-being of our animals. For example, pet owners are notoriously bad at reading (and heeding) behavioral indicators of pain, which can mean that an animal with a painful condition such as osteoarthritis or gum disease will go untreated because the owner doesn't notice that anything is wrong. Many people are under the impression that animals will whine or howl or cry out if they are in pain. And they will. But whining or crying or howling is often evidence of pain that has escalated to intolerable levels. Being able to identify subtle behavioral messages is also important in allowing an owner to step in and intercept a social exchange (e.g., between a dog and a child or between two dogs) that could get dangerous.

It is particularly important to be clear communicators when it comes to teaching our animals the skills they need to thrive in human environments—both inside our homes and, especially for our dogs, when they go out into the community. As I know from hard personal experience, being an effective communicator is challenging. The more books I read on dog behavior, the more I realize how ambiguous my signals tend to be. One of my favorite training guides is the late Sophia Yin's *How to Behave So Your Dog Behaves*, and her title teaches us our first important lesson: 90 percent of dog training is really human training. An example: someone rings the doorbell and the dog barks like crazy. Our first reaction might be to yell "*Shut up!*"—which is like saying to the dog, "Oh my god, what's that? Intruder? Let's keep barking!"

Some veterinary behaviorists distinguish between a well-behaved dog and an obedient dog.[6] An obedient dog does what he is asked, even in situations of conflict or high arousal. Yet an obedient dog might still have annoying behaviors, like attention seeking through constantly pawing his owner. A well-behaved dog, by way of contrast, acts in a "socially appropriate" manner within his human environment. He responds appropriately in diverse situations and doesn't disrupt the owner with his behavior. Creating a well-behaved dog requires an approach to communication and teaching that, according to a chapter from *The Social Dog: Behaviour and Cognition*, "goes beyond behaviourism with its focus on obedience training to one that embraces the more cognitively complex world of dogs in order to develop life skills." Humans and dogs have different communication systems and "different prejudices in processing information extracted from the environment and from each other."[7] This is one reason dogs "misbehave" and ignore our commands. A command typically includes both verbal and nonverbal elements, even if these are not recognized by the owner, and the dog must process both correctly. In turn, the owner has to process and correctly interpret the signals given by the dog. Things can go wrong on both ends of the dyad. If a command has an unrecognized or novel sound quality (e.g., tone of voice), a dog may not obey. So, a dog accustomed to responding to a command given in strong, happy voice may not respond to the same command shouted in anger. Dogs are also sensitive to context, such as physical location. A dog who is trained to "sit" in the kitchen, may not respond to "sit" at the beach or in the dog park.

Our tendency to anthropomorphize dogs can lead to tensions and misunderstandings because a pet owner can draw problematic inferences about why a dog did something or other. It is quite common, for example, to hear a pet owner say, "My Max peed in the house because he was mad that we left him at home all weekend. He was being spiteful." Just because humans are spiteful, we shouldn't assume that dogs are.

Cat-human communication has historically been given less attention in the scientific literature, but this seems to be changing, and there is increasing interest in how we understand our cats and they us. New research is confirming what cat owners have long understood: cats have various ways of telling us what we can do to best serve them today. They speak to us with their soft come-hither meows and with their urgent feed-me-now MEEEOOOOWS. And we learn, at risk of sharp-clawed reprimand, to read their eyes and their tails and the flow of their bodies. Some claim that cats don't really communicate much through facial expressions. Temple Grandin, for example, says in *Animals Make Us Human* that cats have inexpressive "kind of blank" faces.[8] She really ought to meet Thor, whose face is full of emotional information. Paul Leyhausen's classic *Cat Behavior* includes an ethogram of various cat facial expressions and their possible meanings—convincing evidence, to my mind, that cat faces have much to tell us, even if cats are somewhat more elusive subjects than dogs.

The stereotype of cats as aloof and self-involved is gradually eroding, as research into cat behavior intensifies. Perhaps cats are not as overtly empathic as dogs, who seem to know just when we need a head in our lap or a nuzzle on our hand. But cats *are* paying attention. For instance, a recent study on "social referencing" behavior in domestic cats aimed to determine whether cats read the facial expressions and tone of voice of their humans and, particularly, whether cats use this information to guide their behavior. Researchers used a novel stimulus—in this case an electric fan with green ribbons attached—and had the owners regard the fan first with neutral affect, then with happy facial expressions and voice, and finally with fear. More than three-quarters of the cats looked to their humans to see how to respond to the fan and behaved accordingly.

Living with an animal such as a rat, where there is little evolved communicative reciprocity, requires a pet owner to work harder at reading behavioral signals. Behavior is a window into to the emotional world of an animal and can tell us if they feel comfortable and safe or if they are fearful or sick or suffering from pain. Rats communicate with chirps that are mostly inaudible to the human ear, but we can hear them bruxing (a soft grinding together of their incisors, a sign of relaxation) and we can see the eye boggling that often accompanies bruxing, where the eyes seem to jiggle in their sockets. The first time I saw one of my daughter's rats eye boggling I thought the poor animal needed to be rushed to the vet. Luckily, my daughter had read about a hundred books on rat behavior and knew exactly what was going on: the muscle that moves the jaw passes behind the eye, so when the jaw moves back and forth rapidly, as it does with bruxing, the eyeball actually moves, too. She also knew right away if one of her rats was not doing well because he was "puffed out." Piloerection, which causes the hair to stick out, is often a sign of stress or can signal that a rat is in pain.[9]

Given our penchant for talkativeness, it is worth reminding ourselves that to some animals human verbal assaults are not particularly welcome. If, for example, I were to pick up a rat in PetSmart and bring him right up to my face and start cooing "Aren't you a sweet boy! What a cute boy! Look at those cute little whiskers!" I suspect the rat would feel distinctly ill at ease. Similarly for geckos and snakes and mice and hermit crabs, human voices can be quite frightening.

Although talking to our animals, and especially listening to our animals, is a crucial part of building good relationships, it is also sometimes useful to just quiet down a little. When we first adopted Bella from the shelter, and I was trying mightily to teach her how to be a well-adjusted family member and citizen of the human world (and she needed *a lot* of work), I read in one of the training books that it was actually good to try to limit how much you talk to your dog. This was a revelation to me because I had assumed that all friendly interaction is good interaction. The training book explained that if you keep up a constant flow of words, a dog can get desensitized: the sound of your voice simply becomes part of the background. So, when you ask your dog to do something specific, like "come" or "leave it," she may think it is just more of your babbling.

I also wonder whether our animals don't just need some peace and quiet. The soundscape is an important aspect of human well-being, and environmental health studies consistently show negative health effects from living near a loud factory or freeway or airport. The acoustic environment is also vitally important to the welfare and stress levels of nonhuman animals. High levels of noise are known to adversely affect welfare of laboratory animals and can alter behavior and physiology. Unnaturally high levels of noise have impacts on wild animals, too, as seen, for instance, in the deadly effects of sonar on whales and other cetaceans. Noise is a serious problem for dogs and cats in the shelter setting and is a significant factor contributing to so-called kennel stress. I've never seen research on the acoustic environment of pets in the home, but I suspect it is a significant welfare concern. I worry that our goldfish Dibs and Klondike are stressed out by the constant hum of their filtration system. (It is loud enough that I can hear it from downstairs.) We've tried a variety of systems—including the Tetra Whisper and the Aqueon QuietFlow—and they are all equally loud. But the fish can't do without a filter, so what should I do?

Some people emphasize forms of "knowing" and "speaking" that are more metaphysical or spiritual and that cannot so easily be put into words or translated into data. Think of J. Allen Boone's popular 1970s classic *Kinship with All Life*. The modern iteration of J. Allen Boone is the animal communicator. You can actually look in the yellow pages or search the Internet to find one in your area. These people claim to be able to talk to our pets using telepathy or mind-to-mind communication. If your dog has gotten lost or your cat is ill and you want to know when she is "ready" for euthanasia, you can enlist the help of a communicator. After a two-minute search online, I had a list of seven communicators within a twenty-mile radius of my home, and I'm sure I could have found more if I had kept looking. I live close to Boulder, Colorado, which is likely to have an unusually high population of communicators.

I was pretty resistant when I first sat down with one of the many books on psychic communication with animals. But I resolved to approach the material with an open mind. I chose Marta Williams's *Ask Your Animal*, a guide to "resolving behavioral issues through intuitive communication." She says anyone can communicate telepathically with his or her pet—you don't have to be blessed with psychic abilities, though it certainly helps. All you have to do is set your intention, and "think" the message that you want to send. Then you open your heart/mind and "listen" for the answer.

It was hard reading for me, but not because I resist the idea that we can communicate without or, perhaps more accurately, "beyond" the usual channels of verbal and nonverbal information. What I found problematic, in Williams's work, was the idea that dogs and cats and other animals would talk to us in human words, that the telepathic communications would be in English (or French or whatnot). She writes, "You can receive whole words or phrases, or whole sentences and paragraphs, mentally" from your pet, like the cat who told Williams to look for her "up high." (He was on the neighbor's roof.)[10]

I tried what Williams suggested. Bella was asleep on the couch next to me. I set my intention and thought of the words flying through the air from me into Bella's head, "Do you want to play Frisbee?" To my delight, Bella immediately lifted her head and grabbed the orange Chuck-it Flying Squirrel lying next to her. She jumped off the couch and looked at me with that special oh-my-god-it-is-time-to-play twinkle in her eye. I'm still not convinced, but neither am I unconvinced.

What seems most valuable about intuitive communication is the strong focus on the human-animal bond and the recognition that openness and calmness will help people understand animals and will help them to be understood by us. What I worry about is that psychic openness, if not coupled with a practical knowledge of animal behavior, might leave our animals without the kind of help they need. One of the common uses of animal communicators is during end-of-life care, where they might be called in to help a family know when Teddy the terrier is ready to let go of life and, presumably, be euthanized. What Teddy may really need in this situation (in my opinion) is a veterinarian who is skilled at identifying and addressing signs of pain and suffering.

One of the most transformative aspects of sharing our lives with companion animals can be the sense of bridging a gap, of learning how to communicate with each other, of the blending of selves, or what philosophers call intersubjectivity. This journey of discovery can take many forms, from learning about the natural history and behavior of a particular species to becoming familiar with the patterns and personality of an individual animal—perhaps even becoming open to new kinds of kinship.

11. Animal Bling

She walks in beauty, like the night.

Lord Byron, "She Walks in Beauty"

Donna, a colleague of my husband's, has a little Westie named Tinkerbelle. Donna admitted to me recently, rather sheepishly, that she loves to dress Tink up in cute little outfits. "Tink always gets excited when I hold up an outfit for her to put on," Donna told me. "She loves to look pretty." When I asked my friend Sandy whether she ever dressed up her dog Punch, she laughed incredulously and said, "Of course! Especially on Broncos game days." I admit to dressing my dogs in Halloween costumes when my daughter was young because she thought it would be fun. Maya was a witch. We laughed, and took pictures, and said, "Oh my god, how cute!" And Maya got subdued and put her tail down a little. She was still happy to go trick-or-treating with us but would undoubtedly have been much happier in just her own skin. My daughter once decorated Maya's nails with sparkling pink polish (she looked ravishing!). Our animals are an expression of us—especially our dogs, since they go out into the world with us far more than most other kinds of pets. And it is fun to make them fancy, and it's mostly harmless and perhaps even enjoyable for them, at least if they enjoy basking in the glow of our attentions.

Here is a quick sampling of some new "animal bling" items for sale: blow-art pens that you can use to draw on your pet's fur (nontoxic), glittery nail polish (also nontoxic), handwoven sweaters, lavender-scented body wash, and embroidered collars. A company called Rear Gear ("No more Mr. Brown Eye") is selling butt covers for dogs and cats. These little patches "cover your pet's rear while boosting his confidence." You can choose from various styles, including the hazardous waste symbol, a sheriff's badge, a heart, and a cupcake. These products may sound silly, but they are likely also harmless.

We can most certainly go overboard with animal bling. And the examples of extreme overdoing are easy to find, such as real fur coats for dogs, which seem wrong on many levels. But some of the weird products people dream up for pets actually improve the welfare of our companions, like Snoutstik for dry dog noses and silicone nonslip nail covers to improve traction for wobbly elderly dogs.

Some animal bling, however, crosses a line because it has a negative impact on the welfare of the animals. An example of questionable "decoration" is pet body art such as tattoos and piercings. I'm not talking about the tiny tattoos that are increasingly being used as a permanent form of identification or to show that an animal has been spayed (Bella has a little green dot on her tummy, put there by the shelter). These are useful and worth momentary discomfort to an animal. I'm talking about the man who had "Alex/Mel" in a big red heart tattooed on his dog's side and the woman selling "gothic" kittens with pierced ears over the Internet. In response to cases of this sort, several states have passed legislation making it an animal cruelty violation to tattoo or pierce an animal.

Animal bling is also problematic if pet owners blow their pet budget on unnecessary extras while failing to spend money on essentials. I know a woman who buys several new outfits for her dog every month, yet won't take her dog to the vet because "she can't afford it." Is she an anomaly? It appears, from industry sales figures, that pet owners spend more annually on pet supplies (not including food) and grooming than they do on veterinary care.[1] Of course, aggregated sales figures don't tell us about individual spending habits. But the figures at least raise the possibility of misplaced priorities. Pet owners, on average, spend far less than they should on veterinary care. At least a quarter of all dogs and cats never see a veterinarian, and millions live with untreated chronic pain or slow-moving illnesses that owners either fail to notice or are too tightfisted to address. Are the people who spend on fancy collars and personalized food bowls the same people who will spend generously at the veterinary clinic to keep their pet healthy and comfortable? I don't know the answer, but I hope it is yes.

The pet industry encourages us to consume appealing products for our pets. But the basic product on which all further spending depends is the live animal. I address the morality of selling animals as commodities in later chapters (see, esp., chap. 39, "A Living Industry"). Here I would like simply to suggest that there are some instances of animal bling in which the animal him- or herself *is* the bling, and this, in my humble opinion, is not okay. An egregious example is the so-called living necklace. A pendant is made from a little plastic pouch filled with water, to which a small creature such as a turtle or fish has been added. Turtle and fish necklaces have become very popular in China. Needless to say, the necklaces can be worn for only a short time because the creatures quickly perish and start to rot. A somewhat more ambiguous case, likely to be found in at a store near you, are brightly decorated hermit crabs. Hermit crab shells, which are normally a rather drab brown, are painted with swirling rainbows, or super bowl team colors, or with a superhero logo. The problem isn't with the painting itself, which may not harm the crab. But it may change the nature of the consumer purchase. American shopping malls now often have booths called The Crab Shack or some such, which encourage impulse buying by displaying the painted crabs as if they were toys. The crabs are sold inexpensively with a small plastic "habitat" that measures about the size of an index card. Here is where the hapless crab will likely spend the remainder of his short and lonely life. Hermit crabs, in case you don't know, are social animals.

12. Butt of the Joke

Rear Gear butt covers play on the issue of shame. The advertisements suggest (playfully, I think) that dogs are embarrassed by "Mr. Brown Eye" and would prefer to keep their butts discreetly covered. Not surprisingly, Rear Gear has attracted considerable attention in social media, providing antidog people a great opportunity to point out how ridiculous dog owners are. Rear Gear has also raised the ire of some dog owners, who claim that it is humiliating to make a dog wear a "rear enhancer." Silly as it may be, Rear Gear does point to more serious questions. There is a thin line between enjoying the cuteness of animals and getting pleasure from humiliating them, between laughing *with* them and laughing *at* them. When we put animals in situations where the amusement revolves around humiliation, this is no longer harmless fun. Whether animals have the capacity to feel shame or humiliation is beside the point: what matters is our attitude and intention toward them.

Humiliation is particularly serious when the welfare of an animal is put at risk. Take, for example, a group of humans having a party and someone decides it would be funny to get the dog drunk. Since many dogs enjoy the taste of beer, it seems the dog willingly participates in the escapade. Pictures of the beer-drinking dog are snapped and posted on Instagram or Facebook. Everybody laughs. This isn't quite as harmless as a Halloween costume, however, and verges on abuse. Consider the recent brouhaha over a couple of fraternity boys who photographed a young black lab named Mya doing a keg stand—she was held upside down over the keg, with the beer tap in her mouth—and posted it on Twitter. The boys are facing animal cruelty charges.

This is as good a place as any to mention a recent Internet phenomenon called "pet shaming." These photos and videos show animals looking guilty, offering written confessions of their sins, often surrounded by damning evidence. Most of the shamed pets are dogs and cats, but I've seen a few pictures of shamed chickens and a shamed turtle. Maymo the yellow beagle admits that she ate the face off a stuffed elephant; Shimano the pug says, "I puked on the bed 3x and ate it before mom was able to clean it up." A pretty calico cat admits, "I peed on the hamster," and a black cat is sorry to have eaten the $400 tax refund check. Pet shaming makes me feel good because I see that other people's pets are just as mischievous as mine and maybe even worse. But especially I like that these pet owners are good-natured about their pet's "misbehavior."

13. Planting Seeds of Empathy

My friend Maggi received the following note from her eight-year-old daughter:

> Dear Mom,
>
> I am wanting a hamster for my birthday. My stand is they are really really CUTE!!! Also, I will be really responsible.
>
> First, I will take really good care of her. I will feed her everyday. Also, I will play around and teach her tricks, walks around my room, and going on runs in her running wheel.
>
> Second, I will clean her cage every Thursday in a week by putting new litter in. Getting new food in her bowels and in her bottle. Also, putting in her toys such as: tunnels, running wheel.
>
> Everybody will love her when they come over to the house. Also, I will let her run around my room for exercise. I will also let her crawl on my friends and I.
>
> In conclusion, we should get a hamster because of all those details I just listed. Also, I think it would be good for me. I am responsible enough for one, don't you think?
>
> Love,
> your daughter

This will sound familiar to many a parent: the child wanting—no, *needing*—a pet, willing to do whatever it takes to get one. Should we give in to their pleading?

A few weeks ago, I watched unfold a simple scene that speaks to the depth of our animal attraction. I was in Discount Tire, waiting in line for a tire rotation. In front of me was a women holding a baby girl maybe eight months of age. A woman with a Pekingese dog came in and joined the line behind me. The moment the Pekingese walked through the door, the baby's eyes were glued to the little dog. I love babies and tried to weasel my way into her attention by smiling and cocking my head. But she had no interest in me. It was all about the dog. And it shouldn't have surprised me. I had just been reading Gail Melson's wonderful book *Why the Wild Things Are*, which explores the role of animals in the lives of children.

The human fascination with animals begins almost immediately on birth and is deeply rooted in our biology. Infants process visual images of animals faster than images of inanimate objects. Nearly all young children have an innate curiosity about and interest in animals and the development of children into adults inevitably involves the formation and acquisition of a set of beliefs about and attitudes toward animals.[1] For many children, these developing relations toward animals are shaped by the presence of pets in the home. People who grow up with an animal companion are far more likely to see living with animals as an important part of their life; they become the pet owners of tomorrow. Children who grow up with dogs are likely to become dog owners; cat children become cat owners.

Ethologist Konrad Lorenz wrote that childhood pet keeping provides "all we need in order to plant in human hearts the joy to be found in creation and in its beauty."[2] Was Lorenz right? Does childhood pet keeping instill in our young ones a sense of connection to nature, a compassion for animals, and a sense of responsibility? Or does it teach our children that animals are here, above all else, for our whim and pleasure and that an appropriate way to treat an animal is to lock her up in a cage, to deny her any life of her own? Are pets good for children? And are children good for pets?

Relatively little research has been done on whether the presence of pets in the home teaches children responsibility.[3] I suppose it depends on exactly what we mean by "responsible." If our goal is to teach children how to responsibly take care of a companion animal, then having them grow up in the company of pets and having their parents mentor responsible care will most certainly help them acquire these skills. The key, here, is that the parents have to mentor good care, not just hand this responsibility over to the children and not, as too often happens, hold the child to a higher level of responsibility than they themselves are willing to accept. If the goal is to teach children responsibility more broadly—such as taking responsibility for getting one's homework done on time or pitching in to do one's fair share without being asked—then I doubt that caring for a pet is any more like to teach these skills than, say, doing homework or pitching in.

Growing up in the presence of pets does, according to one recent study, seem to have a positive correlation with community involvement and social responsibility. Research by developmental psychologist Megan Mueller found that it wasn't the presence of an animal in the house, per se, that influenced children, but the quality of the child-animal relationship. The more involved they were in caring for the animal, the more likely to engage in community service and help friends and family. According to Mueller, "The young adults in the study who had strong attachment to pets reported feeling more connected to their communities and relationships."[4]

Developing empathic skills is one of the main reasons parents cite for getting their child a pet. This was my main motivation with my daughter. I wanted her to develop what German philosopher Arthur Schopenhauer called "boundless compassion for all living things." According to James Serpell and Elizabeth Paul, pets can act as animal ambassadors, as a kind of "moral link with other categories of animals" and with nature. Pets are able to do this because of their intermediate position between human and animal. And, in fact, some research suggests a correlation between positive relationships with pets and the development of humane attitudes toward animals in general, as well as the expansion of empathy for other people. One study, for example, found that, in preschool children, empathy for companion animals is correlated with empathy toward other school children. Children with strongest bonds to animals showed the highest empathy scores.[5]

Children who grow up with pets may also become more empathic adults—particularly toward animals. Serpell and Paul found that childhood pet ownership was strongly positively correlated with concern for animals in general and with ethical food avoidance (vegetarianism), membership in animal welfare organizations, and, though more weakly, with membership in conservation and environmental organizations.[6] Empathy is not just an emotional response: it is informed by knowledge and experience. The greater the understanding of the other, the greater the potential for empathy. Living with pets, children may learn to understand animals, and particularly animal emotions, in ways that children without pets do not.

But the "pets build empathy" research is hardly conclusive. We don't yet have a good handle on exactly how to study empathy toward animals and humans and how pet keeping influences child development in this area. Some studies show that children with pets have no greater empathy than children without them. It seems intuitive that empathy is a broad, generalized trait, so that empathy for other people and empathy for animals would be linked (the more you have of one, the more you have of the other). But we don't have research that supports this intuition, nor do we know how animal-directed and human-directed empathy are related to childhood pet keeping.[7] Furthermore, empathy can be narrowly focused, so that a child might develop empathy toward dogs and cats yet remain unmoved by the plight of pigs, chickens, or insects. We send our children ambiguous and ambivalent messages about animals and compassion, and having pets could simply add to their confusion.

Another motivation of parents in allowing children to keep pet animals is to engender an interest in science, particularly in biology. This was another of my central reasons for indulging my daughter's animal interests and encouraging her to keep a variety of different creatures. And in this I was successful: she spent her childhood reading books about rats and geckos rather than watching *Hannah Montana*. And now, as a high-school student, biology is her best and favorite subject. Research on how childhood pet keeping influences attitudes toward science and interest in the natural world is lacking, but I suspect that growing up with animals does, indeed, encourage a sense of scientific curiosity in many children.

Now, on to the second part of our pets and children question—the part that has received little attention in the scholarly literature and that is rarely broached in polite company: Is it ethical to put the welfare of a sentient creature into the hands of a not-yet-mature, still-developing little person? If you have had children and pets, I suspect that this question might make you a little uncomfortable. It does me. As much as we want our children to become responsible adults, they are, in fact, children.

It seems unfair to use an animal to teach responsibility when this means that some of the time—those times that the child hasn't quite mastered the art of responsible care—an animal will be left high and dry. I know one family in which the children had responsibility for feeding the dog. When they forgot, the dog went hungry. Is this a case of bad parenting or negligent care for an animal or both? Either way, the dog is the one who suffered. Another friend did a test to see how much commitment her four girls had to their pet rats. My friend put the rats' cage in a closet, out of obvious sight (she was still feeding them and checking their water). It took the four girls, aged eight to sixteen, two weeks to notice that the rats had disappeared. The takeaway lesson for the kids was useful ("you shouldn't have pets if you aren't really all that committed to it"), but in the meantime, the rats had two weeks of sitting in a dark closet.

What children really want to do, of course, is touch and hold.

Yet children are often rough in their handling of animals. Younger children, especially, don't yet have good fine motor control and may actually have a difficult time being gentle to an animal, as hard as they might try. Often they simply haven't been taught how to gently hold a small animal or how to properly show love to a dog or cat (e.g., *not* by squeezing as tightly as possible around the neck). Ironically, the pets that parents often choose for children are the pocket pets, since they are cheap, short-lived, easy to care for (if you simply do the bare minimum), not demanding in the way a dog or cat can be, and easy to give away when the child gets bored. Being so small, however, these animals are far more delicate than a dog or cat and more likely to suffer from rough handling. They are also less well adapted to human environments and more easily stressed by loud noises and lots of activity. The animals typically sold by pet stores as "starter pets" for children—the hamsters, gerbils, hermit crabs—tend to be more hardy (that is to say, more difficult to kill) than "advanced pets" such as the ball python or sugar glider. But "starter pet" is a misnomer. Leopard geckos, for example, are usually placed in the starter pet category, but their needs are quite complex, and providing an adequate environment is challenging, even to those knowledgeable about geckos. Moreover, geckos do not generally like to be handled.

Children and animals playing together is a heart-warming sight, and one of the most compelling reasons for having an animal in the house is to provide a playmate for a child. But often, children engage in forms of play that are one-sided and sometimes even distinctly unpleasant for an animal. When I was young, for example, I was fond of dressing up my dog Brownie. He hated it, and part of what was so amusing to me was the grumpy look on his face. Even though it wasn't fun for Brownie, this probably doesn't constitute abuse. But some forms of play can verge on being downright cruel. I'll never forget the day I went to a friend's house and found her daughter in the bathroom. The bathtub was full of water, and this little girl had her two pet rats in the water, having "swimming practice." If you've read descriptions of the so-called forced swim test used to study the effects of profound stress and hopelessness on rat physiology, this is exactly what it looks like—rats in a tub full of water, with no way to get out or even rest because the sides are too high and too slippery. Where was the girl's mother in all of this?

We also simply cannot ignore the data on abuse of animals by children, which seems to be a distressingly common "developmental" experience (see chap. 26, "Cruelty, Abuse, Neglect"). Most childhood abuse is directed at the family pet, not at strays or lost animals who happen to wander where they shouldn't. Sometimes children learn abusive patterns from their parents, but not always. Having nice parents is no guarantee that a child won't mistreat an animal. When pets and children occupy the same private space, animals are exposed to a certain amount of risk and potential for harm.

The presence of an animal in the home can also provide children an object lesson in mistreatment, ambivalence, even abuse. Parents who are cruel to animals mentor their children in cruelty; violence toward animals is often passed on from one generation to another, like a hereditary disease. Pets also frequently get caught in the crosshairs of family violence and are used by parents to punish or intimidate a child (see chap. 30, "The Links"). Even though the real violence may be directed at the child, the animal becomes collateral damage. One of the most chilling stories of abuse I've ever heard was the mother who was displeased with her son's lack of dedication to his homework. His punishment? She made him kill his hamster with a hammer.

I'm not saying that children and pets are a bad combination. In fact, animals can be profoundly important in the lives of children, particularly in the development of empathy and compassion. One clear avenue toward compassion and respect for animals is the formation of a close friendship with a particular animal. And watching and learning about the biology and behavior of an animal can contribute to the child's developing knowledge about how the biological world works. Having animals in the house also provides the opportunity for parents to teach children how to interact with animals in ways that keep animals and children safe from harm.

Still, there are trade-offs that must be acknowledged, both for the child and especially for animals. And it remains unclear whether certain forms of pet keeping popular with children today—removing animals from their wild homes and making pets out of them (e.g., collecting frogs or tadpoles from a nearby lake) and the keeping of small animals, reptiles, amphibians, and fish in tiny cages or tanks and calling it "good care"—truly teach children compassion.

As for the sweet letter from Maggi's daughter requesting a pet, friends urged caution. "The hamster," they warned, "is a gateway pet."

14. Pets and Our Health

"Animals make us healthy!" is one of the most frequently repeated factoids about pet keeping. Nearly every week some sensational new headline appears: "Why Having a Dog Keeps Kids Asthma-Free" (*Time* magazine), "Your Dog Can Be the Secret to Weight Loss" (Cesar Millan's website); WebMD tells us that "owning a pet can ward off depression, lower blood pressure, and boost immunity. It may even improve your social life."[1] These catchy stories make it sound like having a pet is the best thing you can do for your health and happiness. As with much media reporting on science, such stories typically focus a spotlight on one small research study and take from it the splashiest sound bite that can be inferred (reasonably or not) from the data. They give the impression that we understand far more than we actually do about the intricate network of connections between animals and human health.

The area of most interest, at least of late, has been the various ways in which companion animals can positively influence our health. Although piecemeal, the research on health benefits is intriguing. Here are just a few of these bits and pieces (all of them drawn from Bruce Headley and Markus Grabka's comprehensive 2011 review of the literature).[2] One of the first large studies of pet-health relationships, carried out in the 1970s by Erika Friedmann and her colleagues, looked at the possibility that pets produced physiological changes in their owners and asked whether these changes might be beneficial to our health. Patients admitted to a coronary care unit had better outcomes if they had a pet at home than if not. Other research has suggested that people with dogs are less likely to die in the year following a heart attack than those without dogs. Of a large group of Medicaid enrollees, those who owned pets were less distressed by adversity in their lives and visited the doctor less frequently. Elderly people with pets with whom they are closely bonded declined less rapidly than non–pet owners. Children who grow up in households with pets may, indeed, seem to be at lower risk for developing asthma. The presence of guinea pigs and dogs seems to help encourage social interaction in children with autism spectrum disorders.

There is also evidence that pets benefit our mental and emotional health, an idea that was pioneered by Boris Levinson, who used his dog as a cotherapist. Levinson found that a dog could be a valuable tool for the psychotherapist because the mere presence of a dog facilitated interaction, opened people up, and helped them relax. Animals seem to have a stress-buffering effect, as indicated by various studies measuring physiological parameters such as blood pressure in the presence of pets. For example, in a Japanese study of elderly people, stress levels were substantially lower when walking a pet than when walking alone. Other studies have shown that having a companion dog in the room lowers blood pressure more effectively than an ACE (angiotensin-converting-enzyme) inhibitor. Interacting with dogs can increase the production of oxytocin, a hormone-like peptide associated with feelings of trust, social bonding, and pleasure. Ongoing research is exploring how animals may benefit people suffering from depression, posttraumatic stress disorder, and Alzheimer's disease.

Animals are a form of social capital and can serve as a social lubricant: they provide bridges or links to other potential friends and help people initiate social interactions. And social capital—our network of social relationships and social support systems—is linked to better health.[3] Dogs can help men meet women, and cats can make men appear stylish and sexy. (In a weird gender inversion, even though men with cats are sexy, women with cats are stereotyped as crazy.) Pets can help ease the pain of social rejection and seem to make many people just plain happy. Maybe pets also make us smarter. The presence of a dog during a laboratory-based test reduced the number of errors on a cognitive task and improved retention of information.

Still, caution is in order in how we talk about these various studies. We can't really claim, for example, that "pets lower blood pressure" because the research has not yet systematically addressed different populations, different types of interactions with animals, specific conditions, and so forth. It is also quite possible, for instance, that certain types of families are more likely to have pets than other types and that these same families are also less likely to have children with asthma. Establishing cause and effect is difficult, and most of the research on pets and health should be approached with a degree of skepticism.

Despite the many benefits of living with animals, there are health burdens, too. We are exposed to pathogens, and we can become one of the 4.5 million Americans who sustain a dog bite each year or one of the eighty-five thousand people who suffer a serious injury from tripping or falling over a pet. A perhaps less noticeable but equally serious issue is the stress that pet owners can experience. I think nearly all of my gray hair is courtesy of my pets. Having pets in a society in which not everyone likes animals can lead to social conflict. Even at the dog park, where people share a mutual passion for their canines, I've seen nasty fights erupt among the humans present and have been on the receiving end of some (in my opinion completely uncalled for) impolite remarks.

Dealing with a beloved pet's failing health or impending death can also be a significant source of stress. Ushering my elderly dog Ody through his final year of life was incredibly hard, emotionally and physically. And as I discovered while researching my book *The Last Walk*, I was not alone in my suffering. The pain of losing an animal companion can cripple people with grief, particularly because we live in a culture in which grieving for a pet can feel isolating and socially awkward. If pets really are family, why should it be considered weird to openly grieve their death? And the financial strain of caring for an ill or injured animal can be taxing. Just this month, Maya had an unexpected trip to the oral surgeon ($2,000) and is scheduled for another $1,000 surgery to remove a large lipoma that is starting to affect her biomechanics; Bella has been prescribed three pricey supplements to try to help with her very treatment-resistant allergies; and we've shelled out an additional $350 to troubleshoot Thor's recurring bladder infections. Ouch.

Any decision to bring an animal into the home involves a calculated risk for the pet owner but also involves, even more importantly, taking on responsibility for the life of another creature. Thus far, there has been virtually no research on the health effects of pet-human interactions on the animals themselves. But if we are being careful about animal welfare, this research is vital before we plunge headfirst into pet therapies. How, for instance, do dogs fare in homes with autistic children? Do animals who live in homes with chronically depressed or angry people suffer from emotional contagion?

The "pets are good for us" bandwagon strikes me as too simplistic and irresponsible. Too many splashy headlines pose a danger of making pets into yet another diet pill, açai berry, or fountain of youth—a consumer purchase that might somehow make us thin, beautiful, and young. One recently published book has this to say: "The notion of walking into a pet store and acquiring a pet as a form of self-medication is becoming more widely accepted, even endorsed, by health experts in the mainstream. The idea that caring for and interacting with a companion animal is akin to acquiring a prescription for a combined antianxiety and antidepressant drug is no longer laughable to those concerned with scientific evidence."[4] Animals may be the newest quick fix. But they aren't açai berries. Lives are at stake. Is it really ethical to use a living being as a tool to try to better our own health?

15. Cat Scratch Fever

A class in microbiology changed my view of the world. The public drinking fountain, the doorknob, the refrigerator handle—each had become inhabited by a teeming mass of bacterial and viral predators. Reading through the literature on zoonoses has done something similar: it has made me a little squeamish around my animals, seeing them as hosts to a vast number of microscopic lifeforms, not all of them friendly. I still hug my dogs and cat, but I find myself washing my hands much more often, cleaning up poop more regularly, and turning my head when Thor tries to lick my face.[1]

We are perhaps most familiar with zoonotic diseases originating in livestock or wild animals, such as avian influenza, swine flu, and hantavirus. But the emerging field of companion animal zoonoses is expanding our understanding of these interactions between people and pets. As far as pet-borne diseases are concerned, the one that is familiar to nearly all of us and has perhaps the strongest grip on our collective imagination (especially if you are a fan of *Old Yeller*) is rabies. And rightly so. Rabies infection is fatal. Thanks to aggressive vaccination requirements, rabies is well controlled in the developed countries, though it is still a serious concern in many places around the world. According to the World Health Organization, rabies still occurs in about 150 countries and territories and kills about sixty thousand people a year.

But rabies only scratches the surface of animal-human zoonotic interchange. The recently published *Companion Animal Zoonoses*, by J. Scott Weese and Martha Fulford, is a compendium of information about germs and worms being harbored by our pets. There are the parasites—the roundworms, tapeworms, nematodes, ticks, fleas, and the coccidian parasites that cause cryptosporidiosis, and the protozoa that causes toxoplasmosis. And there are the bacterial diseases, including infection from *Campylobacter*, *Capnocytophaga*, and *Clostridium*, and so-called cat scratch fever caused by the *Bartonella* bacterium. Viral zoonoses, thankfully, are quite rare in the developed world. Finally, we have fungal diseases, which Weese and Fulford say are among the most common pet-associated zoonoses. The routes of zoonotic disease transmission vary widely, but some of the more common ways humans can get infected include bites and scratches, licking or close contact, and contact with feces.

Zoonotic infection from cats and dogs is perhaps of most concern, given the prevalence of these animals in our homes, but other kinds of pets—from rodents, to amphibians, to wild animals—can also carry diseases. We don't generally live in intimate contact with these animals—people don't generally kiss their geckos or sleep with their hamsters (though it probably happens). Still, merely handling these creatures can expose us to pathogens, as can bites and scratches, and these are often the pets that parents choose for children. Bites and scratches from small rodents are fairly common and can make people sick. Rats, for example, can transmit rat-bite fever, which is commonly caused by either *Streptobacillus moniliformis* or *Spirillum minus*. The Centers for Disease Control issued a public warning in 2014 after a ten-year-old boy died from rat-bite fever after getting scratched by his pet rat. Reptiles can carry *Salmonella*. From fish tanks, we can acquire "fish tank granuloma" (*Mycobacterium marinum*). Many interesting connections between pets and human illness are still emerging. For example, there is some evidence linking cat bites with human depression, especially in women.[2]

As Weese and Fulford caution, when we raise the profile of zoonotic diseases and the risk of transmission from animal to human and back again, we also risk causing a backlash against animals. Backlash can come on a broad scale, as has happened in a number of places around the world—for example, in the 2006 roundup and culling (through clubbing, electrocution, and burying alive) of some fifty thousand dogs in Yunnan Province of China in response to a rabies scare. And it can happen on a small scale, when a person decides, often on the advice of a veterinarian or physician and often without evidence, that their pet is making them sick. Weese and Fulford say they have encountered "many unfortunate situations where removal or euthanasia of a pet was recommended based on little to no evidence suggesting the pet was a source of infection."[3] We need, they say, to take a cautious and evidence-based approach and seek an understanding of the delicate balance between the risks and benefits of pet ownership.[4]

16. Pets and Their Health

Having pets in our homes may have health benefits for us. What about for our pets themselves? Do they derive health benefits from living with us? Do we lower their blood pressure, reduce their risk of allergies, and increase their overall longevity? I think the answer is rather mixed and will depend a great deal on the individual pet owner. One thing is for sure: the health of our pets rests almost completely in our hands. If responsible pet ownership means providing for our animals so that their health and longevity are optimal, then conscientious pet owners have a lot of work to do.

Some factors that influence health seem beyond individual control such as inbreeding in dogs (see chap. 36, "Breeding Bad") and exposure to environmental pollutants (see chap. 22, "Pet and Planet"). You could argue, however, that we *do* actually have a measure of control over both of these: we can, for example, adopt mixed-breed dogs and can reduce exposures to many environmental toxins by, for example, not allowing our dog to run around and play on a lawn that has just been sprayed with pesticide. Other factors are clearly important, such as desexing (see chap. 35, "Eunuchs and Virgins") and vaccinations. But the impacts of these on our animal's health are uncertain. In these cases, we are left to make educated choices that we feel balance potential benefits and harms for us and our pets.

And finally, there is a host of factors that we, as pet owners, have almost completely under our control. Nutrition, exercise, and mental health are the Big Three, and these are all covered in other chapters. We also have what might be called "incidental daily care" or some such. This includes those daily or weekly tasks that become part of the routine of care: the brushing of teeth, clipping of nails, unmatting of hair. These may seem trivial but over a lifetime can make an enormous difference in how our animals fare and how they feel. Exactly how much time and effort a pet owner must devote to daily care is up to each individual. And none of us will be perfect, just as we aren't perfect with ourselves or our children. I, for example, do not brush the teeth of each critter—Maya, Bella, and Thor—each and every day, despite the fact that daily cleaning is the gold standard of dental care. The dental care I offer is more like the bronze version: I do it, but inconsistently and not as often as I should.

And then the big one: timely and consistent access to veterinary care. With humans, access to health care services is considered a basic need. We may not agree on how best to secure broad access to medical care, but nearly everyone agrees that doing so is important. Yet with our animals, veterinary care is often approached as a personal preference of the pet owner or a luxury item for our pet, along the lines of a fancy new food dish. The number of cats and dogs who never, in their entire life, visit a veterinarian is estimated at anywhere from about 25–50 percent (with most estimates falling on the higher end). No veterinary care, ever? This is mind-boggling. We have no data on access to veterinary care for pocket pets, reptiles, fish, and exotics, but my educated guess is that around 80–90 percent of these animals never see a veterinarian.

Of those who do see a veterinarian, a large number of animals will meet an untimely end because the costs of keeping them healthy are "too much," whatever this nebulous number happens to be for an individual pet owner. Companion animals are often euthanized in lieu of veterinary treatments, to keep costs down, either when an animal needs acute veterinary attention or is diagnosed with a chronic condition and veterinary bills start adding up. Owners are allowed a great deal of discretion, and although some people might consider you distasteful if you throw your animal under the bus at the first sign of financial pain, it is common enough and quite legal.[1] I would propose the following five principles to guide spending on veterinary care for pets.

1. It is not fair for an animal to live with chronic pain or to have their quality or length of life severely truncated by their owner's failure to provide basic care.
2. Therefore, no one should acquire a pet who doesn't intend to provide consistent veterinary care, and who doesn't have the financial wherewithal to do so.
3. Pet owners on limited budgets should plan ahead and have a designated fund set aside for veterinary care and emergencies.
4. It is not fair to expect a pet owner to suffer financial ruin to care for an animal.

Lest this be read as discriminatory toward poor people, the moral target here is not people with limited resources, but people who place disgustingly little value on the life of an animal. People who have the resources but fail to open their wallets, or whose spending priorities are skewed (who are unwilling to spend $1,000 dollars on a life-saving surgery for their dog but who stop on the way home from the euthanasia appointment to console themselves with a new $1,000 television), should not own pets. It is worth noting, in this regard, that the average amount of money American consumers spend yearly, per capita, on veterinary care is considerably less than they spend for access to pay television. And incidentally, research conducted by sociologist Leslie Irvine on homeless people and their dogs shows a remarkable level of devotion, and the needs of the dog are often placed above the person's. The title of her book is *My Dog Always Eats First*.

17. Feeding Frenzy

The dog's dietary philosophy: "If it falls on the ground, eat it. You can always throw it up later."

Dave Barry

What should a responsible and caring pet owner feed to his animal? This seems like such a simple question, but the answer is surprisingly elusive. What constitutes the best diet, especially for our dogs, has become a veritable circus of conflicting opinions, to which adherents cling with surprising violence. A number of forces push and pull on the thoughtful pet owner: the advice of veterinarians, the marketing of the pet food industry, consumer websites that warn of the dangers of commercial foods, the holistic magazines that insist on home-cooked meals, the admonitions of the raw foodies, and more. As in the realm of human diet, nobody seems to agree on what diet is most healthful or most ethically responsible. Meanwhile, all we really care about is feeding our pets food that they will enjoy, that will keep them healthy, that won't break the bank, and that doesn't cause too much collateral damage. To those of you trying to figure this out, I wish you Godspeed. Here are some of the issues I've struggled with (for simplicity's sake, I'm going to focus primarily on dog food, but many of the same questions apply to cat and other critter food): is commercial food more or less healthy than home-cooked? Can we trust the quality of commercial dog food? Is meat necessary for dogs, and if so, how can I reconcile my moral concerns about meat eating (both the animal suffering involved in the meat industry and environmental impacts of meat production) with feeding my critters?

For a long time, there was no such thing as "dog food." During most of our coevolution, dogs lived off the by-products of human consumption. They ate our scraps and also, according to anthropologists, our feces. John Bradshaw argues that it is the close similarity of the human and canine diets that spurred the evolution of the dog as a companion species.[1] When did this collaboration stop? Why must we make a special trip to a special store to buy food specially manufactured and packaged for our pets? Because of the marketing genius of the pet food industry juggernaut.

According to the Pet Food Institute, "dog food" was invented in the 1860s by an Ohio electrician named James Spratt. As the story goes, while in London Spratt observed a group of dogs being fed leftover biscuits and it came to him—like a flash of electricity to the lightning rods that he was in London trying to sell—that he could make biscuits specially designed for dogs, with a combination of animal and vegetable material. From these humble roots has grown the Leviathan that is the modern pet food industry. As the pet food industry has grown, it has drastically and deliberately altered our way of feeding dogs. Scraps are no longer acceptable. Mary Thurston, in *The Lost History of the Canine Race*, suggests that vilification of "scraps" was a deliberate move by pet food marketers to make their products indispensable.[2] Through brilliant marketing, dog food has become one of the most remarkable success stories in American business, creating demand for an illusion worth $22.5 billion and growing.

But is the pet food industry providing our animals with food that is better than the leftovers from our own kitchens? The most popular dog food brand in 2012 was Cesar Canine Cuisine. Despite its popularity, Cesar gets low ratings by several consumer dog food review sites. For example, Canine Cuisine gets 2.5 out of 5 "paws" on the Dog Food Insider website, which rates various dog food brands according to the quality and safety of their ingredients. Below the rating, you can read a long list of personal comments posted by dog owners whose canine companions had become violently ill after eating Cesar food. Kibbles 'n Bits, whose television commercials are so cute, nevertheless rates an embarrassing one "paw." Yikes.

Why the low ratings? And why the plethora of websites dedicated to reviewing pet foods? Because people are afraid to feed their pets. They are afraid that the food we so lovingly pour into a bowl will make our animals sick—or maybe even kill them. And there are good reasons to be concerned. Over the past decade, several large-scale pet food disasters have been linked to adulteration with a nonfood additive, including the 2007 melamine poisoning that killed and sickened hundreds, perhaps thousands, of pets. Jerky treats from China are currently being pulled off the shelf because dogs who eat them are becoming violently ill. And these are just the worst-case scenarios: at any given time, a handful of recalls are ongoing, usually well beneath the radar of consumer awareness. Dog and cat food has been found to be contaminated with *Salmonella*, *E. coli*, *Clostridium*, *Campylobacter*, and large numbers of people have fallen ill from handling pet food. Kibbles have tested positive for aflatoxin B1 and other mycotoxins that come from grains (particularly from inadequate storage). Beyond the problems with contamination and adulteration, many of the basic ingredients in our pet foods are of such poor quality that pet owners are turning, instead, to homemade foods.

A handful of multinational companies control the majority of the pet food market, not only in the United States but also globally: Mars Petcare (Pedigree, Cesar, Sheba), Nestle Purina (Fancy Feast, Alpo, Friskies, Beneful, One), Del Monte (Meow Mix, Kibbles 'n Bits, Milk Bone, Pupperoni, Pounce), Colgate Palmolive (Hill's Science Diet, Prescription Diets), Proctor and Gamble (Iams, Eukanuba). To give you a sense of the scope of the industry's financial power, Mars and Nestle Purina both have yearly revenue of more than $16 billion. These companies source their ingredients from a small collection of global suppliers, so that most of the dog and cat food in the United States contains essentially the same ingredients. We saw this playing out in the 2007 recall: 180 different brands of pet food all used ingredients supplied by Menu Foods—

ingredients that turned out to be deadly. One of the protein sources used by Menu Foods came from China. Although it was sold under the name "wheat gluten protein," it contained large quantities of melamine. Melamine is a cheap and plentiful industrial chemical used in making plastic products. It also has the useful property of showing up in "guaranteed analysis" tests as protein, and it is far cheaper than natural protein sources such as cow, pig, or fish.

The pet food industries and human food industries are closely linked. As I noted before, dogs have historically survived, even thrived, on human scraps. Industrial pet food follows roughly the same principle: use the leftovers. But today's industrial leftovers are much different than the leftovers of yesteryear. These are not leftovers from a hearty family meal of steak and potatoes but are instead the waste materials and by-products of agribusiness. These are the materials deemed, by the United States Department of Agriculture (USDA), "not suitable for human consumption." These are not just the parts of animals that humans find aesthetically unpleasant, such as the hooves and ears and noses and unborn babies. These are also the yellow plastic ID tags that are left on the cow carcasses after slaughter, the fecal material that drains from the intestines, the rat poison and sawdust, the diseased and dead-on-arrival food animals that the USDA says cannot legally be put into human food products.

There is nothing inherently sinister in trying to capitalize on discards or in making good use of animal parts that might otherwise go to waste. And even animal advocates can be thankful that when animals are sacrificed to the market, their entire bodies are used. But there is something menacing afoot when companion animals suffer ill-health, disease, and early death because we feed them garbage.

And here is one of the key problems with pet feed. Pets are thrown into the larger category of agricultural animals. The ingredients used in pet feeds are essentially the same as those used in the feed of pigs and chickens and cows destined for slaughter. Yet there is a low threshold for what is deemed safe for agricultural animals, since they generally live for only a year or two and so the long-term effects of their feed ingredients are considered unimportant. Feed can include "'material from diseased animals' or 'contamination by filth' or 'contamination by industrial chemicals.'"[3] Cows and pigs fed cardboard and sawdust enhanced with antibiotics and growth hormone won't develop cancer at the age of five because they will never live to be five. Indeed, what is important is quick weight gain, at the cheapest price possible. But we really don't want to have a population of obese dogs and cats who die of cancer well before their time, do we? Yet apparently, that's just what we have.

A growing number of veterinarians, as well as pet owners, link the poor quality of pet food to extraordinarily high rates of cancer and other life-limiting diseases in cats and dogs. Poor-quality food has been linked not only to cancers but also to urinary tract disease, kidney disease, heart disease, and dental disease. And, in case you haven't noticed, our dogs and cats are unnaturally fat (see chap. 19, "Your Dog Is Fat!"). Something is amiss when few of our companion animals actually live out their natural life spans. Food, of course, is only one of many factors that influence health and disease; others include environmental exposures, genetic bad luck, and inbreeding. But food is a critical one. People may balk at the price tag on high-quality pet food, but when compared to the veterinary bill for acute kidney failure or lymphoma treatments, the healthier food is likely to be the more affordable option. Of course, some expensive foods are of no better quality than cheaper foods; large companies, in particular, may source their base ingredients from a common third-party company such as Menu Foods, so the higher-priced kibble or cans actually contain essentially the same raw materials as the cheaper stuff, packaged differently. But some high-quality pet foods actually deserve the title.

The pet food industry seems to have an uncommon level of control over safety and quality standards. The industry is regulated by the Food and Drug Administration (FDA) Center for Veterinary Medicine and the USDA, neither of which is known for being particularly transparent. Strengthening USDA inspections and FDA involvement, particularly on imported ingredients, would help reduce the risk of another melamine disaster. But the FDA and USDA don't actually create the standards for what can and ought to be included in feed products. These rules are established by the American Association of Feed Control Officers, a group of animal feed industry representatives.

Whether or not veterinarians know what they should about pet foods, there is no denying that pet owners can have difficulty finding accurate information. Right now, consumers of pet food products are provided very few facts, and certainly not enough to make informed choices about what we want to feed our companions. Pet food labels are allowed to be misleading: there are minimal standards for what constitutes false advertising. According to regulations of the American Association of Feed Control Officers, for example, pet food labels "may include an unqualified claim." So, "homemade chicken stew" can be anything at all—it doesn't even have to contain chicken. And as absurd as this may sound, one of the few things pet food makers *cannot* claim is that they use high-quality ingredients, even if they do.

As with human foods, it is often hard for a consumer to discern the origin of the ingredients or even get a full accounting of what things are. "By-product" and "meal" for example, are notoriously obscure feedstuffs, as is "animal protein." Even "meat" is indeterminate. Some animal protein materials are fine nutritionally, though not particularly appetizing (the hooves, beaks, innards, heads, skin). Other protein sources such as hydrolyzed hair, spray-dried blood, and dried swine and poultry waste may serve more as cheap filler than anything else. Although pet food manufacturers are required to list the ingredients they put into a product, they do not have to list all the things added by their ingredient suppliers. They might, for example, use "fish meal" to which ethoxyquin (a preservative known to be harmful) has already been added. So unless you make a number of frustrating calls to various company representatives and try, for yourself, to ferret out where the fish meal came from and what it contains, you might well be giving your pet additives that you would rather not.

A persistent concern about pet foods is that one of the protein sources ("protein meal") could actually be the rendered remains of companion animals, most likely from the corpses of dogs and cats euthanized at shelters and then sent (because the corpses have to go somewhere, landfills are overburdened, and cremation is too expensive for the high volume coming out of our shelter system) to the rendering facility. The cannibalism issue is unseemly, but not nearly as serious a concern as the possibility that dog and cat food would then contain pentobarbital, the barbiturate used for euthanasia. It is hard to discern fact from fantasy, but according to a Food and Drug Administration Center for Veterinary Medicine study, low levels of pentobarbital have, indeed, been found in a number of pet foods. However, the FDA has also looked for canine and feline DNA in pet foods and reportedly found none. The FDA/CVM study claims that the pentobarbital must have come from cattle or horse carcasses. Still, cattle and horses are not generally euthanized with sodium pentobarbital, so the presence of this euthanizing agent remains a mystery to be solved.[4]

In response to consumer concerns about cannibalism and pentobarbital contamination, the Center for Veterinary Medicine tried to determine whether the levels of pentobarbital found in pet food would be harmful to pets. They did an eight-week feeding trial, using forty-two beagle puppies, who were fed pentobarbital-laced food. The puppies were then killed (!) and their organs evaluated. Researchers found no obvious harm to the forty-two subjects.[5] Thus, the official FDA position is that our pet food is safe. Yet because our pets will probably eat these foods not for eight weeks, but for eight years or more, the data don't have much bite.

With all of this in mind, should we follow the advice of holistic pet magazines and feed our dogs home-cooked meals, where we control the ingredients? I have asked a number of veterinarians for advice on this question, and I usually get the same answer: commercial food is better because it is "balanced" with the appropriate range of nutrients. I don't like this answer, which is why I keep looking for new veterinarians to ask. I want a veterinarian to tell me that home-cooked is better, because this is where my intuition and the bulk of my reading strongly leads. I don't completely trust the nutritional advice of veterinarians who (like human physicians) are given relatively little training in nutrition and what lectures they are given may have been sponsored by Hill's Science Diet. But I also don't trust myself. If I ask my dogs—who are perhaps the ultimate authority on what dogs should eat—the answer is clear: homemade food is highly palatable and gets five paws.

18. Who Should We Feed to Our Pets?

For some people, choices about which pets to keep often center on what it is these pets need to eat—or, perhaps I should say, *who* these pets need to eat. I am acutely aware of this dilemma every time I shop for dog and cat food. As an animal lover, I am troubled by the unholy industry that breeds, feeds, and slaughters cows, pigs, chickens, turkeys, and all the other species commonly found on the American plate. I choose not to eat animal products, raised my daughter vegetarian, and live in a meat-free home—except for the pet food.

I don't like to buy meat for my pets. It makes me uncomfortable. But I do it. Many people, including some holistic veterinarians, argue that a vegetarian and even a vegan diet can be quite reasonable for dogs and even cats, and testify that animals thrive on a plant-based meal plan. The longest-lived dog on record—twenty-seven years—was a Border collie who allegedly lived on a vegan diet. If we compared a healthy, home-cooked vegetarian diet to crappy commercial kibble, the vegetarian food would likely come out on top, in terms of health risks and benefits. But from my study of the literature, the weight of evidence suggests that my cat and dogs need meat to be healthy. This is particularly true for Thor, because cats are obligate carnivores (they cannot fully digest and get nutrients from plant proteins and need a diet that consists mostly of meat). One benefit of cooking for our dogs and cats at home is that we can control the source of any meat we buy, and can, if we are so inclined, purchase from individuals or companies that pay attention to the welfare of the farmed animals. But even "humanely raised" animals endure a good deal of suffering, and buying their flesh can still be morally uncomfortable.

I must admit to another factor in my decision making: the pleasure my animals take in their meals. For a period of several months, I did make my dogs go vegetarian, and briefly vegan, and I suffered many looks of disappointment and approbation when their bowls were presented to them at meal times.

The food-chain issues don't stop at dogs and cats, and often shape the kinds of choices people make about pets. Even in my pet-crazed years—the years of my daughter's childhood—I drew a line at the kinds of pets I would consider: no animals who need to eat other sentient beings, where *we* would have to orchestrate the death. I refused to even consider any species of snake whose diet would consist of live rodents or even frozen pinkies or pinky-parts (a pinky is a newborn rat or mouse). You'll notice that I am engaging in a classic form of moral self-justification called "distancing." When I buy a package of ground turkey to cook for my dogs, I don't have to participate directly in picking the turkey out of the flock and cutting his neck. So the turkey is easier for me to swallow, morally speaking, than the live rat, who would need to be fed, by me, to a snake. Back then, I didn't put insects into the "sentient" category, but during the course of our gecko keeping, I grew more and more uncomfortable with the fate of the crickets. Each cricket started to feel a bit more like a "who" than an "it."

In considering the various moral facets of pet keeping, what and who our pets eat is a central issue, with deadly implications for animals that fall into the category of "pet food" rather than "pet." The fact that one creature, such as a rat, could be both pet and pet food shows the shifting moral sands on which animals stand.

19. Your Dog Is Fat!

When my vet announced to me that Bella was fat, I was embarrassed and ashamed. How could I have let this happen? "You've got to get her weight down," the vet scolded. "It really isn't good for her." Bella has joined the growing ranks of overweight dogs. We are working hard to get her back to a healthy weight, but it is a struggle.

Did you know that there is an obesity epidemic among our pets that nearly matches, perhaps even exceeds, in scope and seriousness the same epidemic among people? According to the Association for Pet Obesity Prevention, more than half of all dogs and cats in U.S. households are obese. More than half! While talking with my vet about Bella's diet plan, I asked her how many of her clients she has to counsel about an overweight pet. "About 80 percent," she said with a sigh.

Most of our pets' weight problems stem from consuming too many calories and not burning enough through physical activity, the same nettlesome calculus that haunts humans. Yet as with humans, obesity is not quite that simple. Excess weight gain can also stem from certain medical conditions or from taking certain medications; genetics can play a role, as can gender and age; and some breeds are more prone to obesity than others. Though not yet widely recognized by pet owners, stress, boredom, and other negative emotional states can also drive animals into unhealthy eating patterns.

Being overweight compromises the health of our pets in various ways, most notably in increased rates of arthritis and other joint problems and in greater likelihood of developing diabetes, heart disease, and cancer. It also, and perhaps even more importantly, compromises daily quality of life for our pets, decreasing their vitality and emotional well-being. It isn't just dogs and cats who become overweight. Small pets such as hamsters, rats, and guinea pigs can easily get fat, particularly since they often spend long hours in a small cage. Even pet fish can get fat. There are various Internet forums on overweight fish, which caution that sometimes a "fat" fish is actually swollen or has a tumor or, perhaps, is pregnant. The basic takeaway can be applied to our critters: if your pet looks unusually rotund, please investigate and perhaps enlist the help of a veterinarian.

Once overweight, it can be quite difficult for an animal to shed the pounds. Many pet owners struggling with weight issues drew inspiration from the weight-loss struggles of a cute big-little dachshund named Obie, via his human owner Nora Vanatta. Obie had belonged to an elderly couple who couldn't give him much exercise but gave him plenty of love in the form of food. When Vanatta adopted Obie, he weighed seventy pounds and could barely walk. (Average dachshund weight should be around twenty pounds, tops.) Vanatta immediately set out to help Obie improve his health. She also, through social media outlets like Obie's own Facebook page, called *Biggest Loser, Doxie Edition*, made the dog a national celebrity and poster child for the obese pet. Obie was able to lose nearly fifty pounds and he looks great.

Now that there are so many fat dogs, thanks in part to excellent marketing of rich foods and irresistible treats, the pet industry has stepped in with various weight-management products. Some of these can be useful, such as specifically formulated diets that can help with weight control. Others, such as Slentrol, the "first prescription weight-loss medication for dogs," offer a quick fix in pill form instead of getting at the underlying behaviors that drive weight gain. As with humans, the best weight control program may be the simplest (and cheapest): smaller portions of nutritionally dense foods, less junk food and fewer empty calories, and more physical activity. Of course, our dogs and cats will not make this easy for us. They will look at us with beseeching eyes and will whine or meow. We, and they, need to have willpower. (The website of the Association for Pet Obesity Prevention has information on determining a healthy weight for an animal and helping reach this goal [http://www.petobesityprevention.com/].)

Weight-loss strategies for pets focus on the physical aspects of weight loss—the number of calories eaten and the minutes spent walking or playing ball. What you rarely hear discussed, either by pet owners or even in the veterinary literature, is the role of psychological factors in overeating and in the development of eating disorders in companion animals.

Veterinarian Frank McMillan did a careful review of the veterinary literature on stress-induced and emotional eating in companion animals. Like humans, animals appear to engage in what we refer to in our own species as emotional eating, spurred by stress, anxiety, or other negative emotional states.[1] He found that although there is a huge body of literature on companion animal obesity, there is relatively little discussion of the psychological aspects of overeating and to what extent stress-related or emotional eating is a contributing cause of the disorder. Some of the few studies that have been conducted suggest an important role for emotion. For example, several researchers have suggested that cats may overeat as a coping mechanism to deal with boredom, anxiety, and depression. One study suggests that apartment-dwelling cats are at increased risk of obesity, presumably because of stress related to boredom.

Overfeeding is not the root problem—overeating is. Before asking how to help an animal lose weight, we need to be asking why the animal overeats in the first place. We have ample experimental data demonstrating that if a group of animals is given all the food they want, even a diet high in fat, some will become overweight and some will not. If an animal's obesity is the result of stress-induced eating, the standard approach to weight loss—restricting the availability of food—may actually make things worse by increasing the level of stress experienced by the animal, feeding a vicious cycle of unhappiness. Being overweight diminishes an animal's quality of life, and McMillan points out that diminished quality of life may be an important contributing factor *to* an animal's weight gain.

Many people reacted to Obie's severe weight crisis by hurling abuses (via Obie's Facebook page) at his elderly owners. While verbally abusing an elderly couple seems a bit misdirected, the question certainly does linger in the air: isn't allowing an animal to become overweight a form of mistreatment? According to a judge in Ottawa, in the case of a morbidly obese cat named Napoleon, the answer is yes. A woman was charged with permitting distress and failing to provide adequate care to Napoleon, who was brought to the Ottawa Humane Society unable even to stand up or groom himself because he was so fat. Napoleon weighed in at twenty-five pounds (instead of about ten), and was so morbidly obese that the shelter staff judged euthanasia to be in his best interests. Napoleon's owner couldn't claim ignorance because she had been counseled to keep his weight down but hadn't listened to the vet's advice. As a result of her conviction, she will have to pay a restitution fine to the Ottawa Humane Society and will not be allowed to own an animal for the next five years.

Of course there is a big difference between Napoleon's situation and the cat who is a pound or two over his ideal weight. And as a huge number of people with animals know, maintaining an ideal weight is a constant struggle, and some animals, like some people, have more trouble than others. For Bella, I'm cutting way back on the number of treats and substituting baby carrots for biscuits. I'm getting more creative in thwarting her counter-surfing excursions and keeping the cat's dish well out of reach. I'm using a measuring cup when I feed her (because I've discovered, after prodding from my vet, that I don't "eye-ball" a half cup very accurately) and, to Bella's great pleasure, I'm adding some extra Frisbee and ball-chase sessions every day.

20. Poop

At least once a week, someone writes into my local newspaper with a complaint about dogs. Here is a recent and quite standard entry: "I was walking at Roosevelt Park today and I saw a woman walking her dog, and after the dog pooped, she looked around to see if anyone was watching, and then SHE JUST WALKED AWAY!!" The complainant then went on to explain how all the good and innocent citizens of our town are in danger of stepping on the poop, getting some dread disease from it, and basically having their day ruined.

Dog feces can certainly stir up a certain amount of animosity toward dogs and dog owners. Just last week, a Texas man shot and killed his neighbor's dog for pooping on his lawn. That's how upset people get. My dog Ody had a sixth sense when it came to knowing which lawns belonged to antidog people. These were where he chose to do his business, and sometimes no dragging on the leash could persuade him to move along to a more private location. This, I will tell you, can create some uncomfortable social situations, and they can get particularly dicey if you discover that the newspaper bag you had in your pocket has fallen out somewhere on your walk and you are left with only your bare hands to clean up the offending pile.

With an estimated eighty million dogs in the United States, the sheer quantity of feces produced daily poses a significant environmental concern (as noted in chap. 22, "Pet and Planet"). Dog feces can also contain potentially pathogenic microorganisms, which can transmit diseases to humans, to other dogs, and to wildlife and livestock (see chap. 14, "Pets and Our Health"). The larvae of the roundworm *Toxocara canis* can, if ingested (through, e.g., contaminated soil on vegetables), cause a nasty illness that can result in blindness, as well as in rheumatic, neurological, and asthmatic symptoms. Recent research has raised concern that dog feces may also contain antibiotic-resistant bacteria.[1]

There is a well-developed literature on the environmental and public health aspects of dog waste and an emerging interest in the social and emotional dynamics of poop clean-up. It may surprise you to know that dog poop scooping has also been the subject of scientific study. After interviewing a number of dog walkers about their attitudes toward poop pickup, researchers came up with the following typology:

Proud to pick up (and proud to be seen carrying dog waste bag; will take it all the way home if bin is not available)
It is the right thing to do (but don't like it, and find carrying poo bags embarrassing; won't pick up if in a remote location)
I have done my job (picks up reluctantly; may ditch bag behind a tree if no bin available)
Only if I have to (picks up when being watched by others; will definitely ditch bag when no one is looking)
Disengaged (not a chance)[2]

Which one of these are you?

Cat poop is not as provocative a social issue as dog poop, since cats generally go in litter boxes or otherwise bury their waste in inconspicuous places. But neighborly disputes still arise. One of our friends, for example, moved into a home that had gravel in the front yard. He soon discovered that his yard was the litterbox for most of the outdoor cats in the neighborhood, and he was not pleased. To give another example, we built a small sandbox in the backyard of our house when our daughter was young, and this, too, became a favored spot for the neighborhood cats. It isn't nice to think of your young child digging around amid cat feces. Indoor litterboxes are a marvelous invention, but I try not to think too hard about the fact that Thor walks around on our furniture after visiting his box. Cat feces can carry various pathogens, including the protozoan that can cause toxoplasmosis, a.k.a. mad cat disease.

21. Animals Bite Back

Birds can peck, hamsters and gerbils can inflict painful bites, rabbits can scratch with their powerful back legs. With a single innocent swipe, a cat can permanently damage a child's eye, and a dog can kill. Sharing our intimate spaces with animals involves a degree of risk. If we have children and we decide to also live with animals, we must accept that we are putting our children at a certain degree of risk, too. The risk can be minimized and managed, but it will always be there. In my own pet keeping saga, I have (in my own estimation, as an older and wiser person) grossly underestimated the risk I exposed myself and my family to, and were I able to go back in time, there are many things I would do differently.

According to data from the Centers for Disease Control, about 4.5 million people get bitten each year by dogs in the United States, and of these about nine hundred thousand require medical attention. In 2012, more than twenty-seven thousand people underwent reconstructive surgery after being bitten by a dog.[1] Thirty people are killed, on average, by dogs in the United States each year. The majority of dog bites and about 70 percent of fatal attacks involve children under the age of nine. Over the past three decades, as the population of dogs has increased, so have the bites. Chained dogs are more likely to attack than unchained dogs, and intact male dogs are responsible for the lion's share of the fatal attacks. Any breed of dog can inflict a bite, but some dogs may be more dangerous than others by virtue of their size and temperament.

One of the most contentious issues within the realm of dog bites is whether certain breeds of dog are inherently more dangerous than others, with the usual contender for "most dangerous breed" being the pit bull terrier. Whether this assignation is deserved is a point of heated debate and statistic slinging. So, too, is the question of how to respond fairly and appropriately to an increased risk posed by certain dogs—if indeed such increased risk can be firmly substantiated. Unfortunately, many policies aimed to reduce perceived risk, such as breed-specific regulations, also wind up causing problems for the entire population of dogs who fit a certain physical profile.

Shelters around the country are full to the brim with pit-bullish dogs. Walking through my local shelter, for example, you would see kennel after kennel occupied by a dog who you might reasonably guess was a pit bull. But you would probably be wrong, were genetic analysis of the dog actually conducted. Shelter personnel frequently misidentify breeds, and, in light of this, our shelter (and others around the country) have instituted a new policy: they will no longer identify dogs by breed—acknowledging that we very often get things wrong and that dogs who are labeled as pit bull may have a harder time getting adopted and may be more likely, in those shelters that haven't embraced no-kill policies, to face the needle. Moreover, "pit bull" isn't even a breed, per se, but a catchall term used to describe a nebulous and continually evolving group of dogs.[2] Like *Minority Report* for dogs, we are judging dogs guilty in advance of any crime. And we're judging them as a group, rather than on their own merits. It seems far better to evaluate the potential risk of each dog as a unique individual.

The statistics do not tell us that dogs are inherently dangerous or that dogs hate children. What they do tell us is that being around dogs places us and our children at some risk. The biggest risk factor for dog bites is having a dog in the home, and with more than one dog in the home, risk increases substantially. When children are involved, risk also increases because children are not as adept as adults at interpreting a dog's behavior. In one study, for example, four- and six-year-old children were not able to successfully identify a fearful dog, although they did pretty well at recognizing aggression and friendliness.[3] The children reported looking at the dog's facial features but not paying attention to posture or tail position, both of which can help us identify fear. Children are likely to misinterpret dog facial expressions and will, for example, as mentioned in chapter 10, interpret teeth baring in a dog as laughing.[4] Children often move quickly and unexpectedly and don't always respect personal boundaries. Although "never leave a child and a dog alone" is common advice, it is often ignored by parents, sometimes with serious consequences for dog or child or both.[5]

Although not as high profile as dog bites, cat bites are also a significant cause of injury. About four hundred thousand such bites are reported each year. Because cat bites often cause deep puncture wounds, and cats' mouths have a lot of bacteria, bites often get infected (by some estimates, 80 percent of bites). Adult women are the most common victim of cat bites. Like dogs, cats generally bite or scratch in response to provocation (such as being given a bath), though many bites and scratches occur during play sessions that turn rough. Many cat owners will also be familiar with a pattern of feline behavior that might be called "petting aggression": your sweet cat climbs into your lap and nuzzles your face until you start petting her. She is purring contentedly when all of a sudden she reaches around and takes your arm between her claws and starts biting.

Although the pocket pets seem small and harmless, these animals will use whatever defenses they can muster, if they feel threatened. My daughter was bitten by a hamster hard enough to require six stitches. Gerbils and mice and rabbits can also inflict nasty bites—they all have long, sharp incisors that will easily pierce through skin. Several children have died of rat bite fever in the past few years after being bitten by pet rats.

Unfortunately, when a human gets hurt, it is often the animal who gets blamed. This is particularly true in the case of dog bites. We often speak of them with an air of disbelief: how can these creatures, who we care for so unselfishly, be so disloyal and turn on us? But dogs express themselves with their mouths and teeth. Not only that, but if we look at the circumstances surrounding dog bites, we will see that these "attacks" are rarely unprovoked. Humans don't always behave nicely around dogs: we taunt them, corner them, make them nervous, do aggressive things like stare at them or try to touch their heads, and generally provide all sorts of triggers for biting.

Lack of appropriate supervision or socialization is often a factor. For example, dogs who are allowed to run loose, whether on purpose or through carelessness, can pose a danger, as can dogs who are untrained, unsocialized, or anxious. Being labeled a "biter" is often a death sentence for a dog.

Of course, there are ways to control risk. The safest course is simply to forgo having a pet altogether or to choose something truly innocuous, like a Chia pet. We can decide against purchasing a wolf, a python, or a baby cougar and opt, instead, for a cat or dog. Children can be taught how to approach an animal, how to read an animal's behavior, and why it is important to respect an animal's personal space and heed warnings of the "time to leave me alone" variety. And parents can exercise caution in leaving children and animals together unsupervised, for the safety both of the child and of the animal.

22. Pet and Planet

As we reach the end of our first section, "Living with Spot," it is fitting that the final consideration is one with global reach. Can you be a good environmentalist and still own pets? Is pet keeping a luxury that we can really afford, in a world of scarce resources, heavy toxin burdens, and global warming? Actually, research has suggested some correlation between pet ownership—particularly dog ownership—and self-identity as an environmentalist. But I have found being Green doesn't always sit comfortably with being a pet owner.

The first issue to consider is the most local in scope: How does the quality of the environment impact the health of our pets? The things that people worry about—the quality of our water, indoor air pollution, pesticides and genetically modified organisms in our foods, smog, fracking wells next door—also threaten the health of our pets. We don't have shelves full of books and reports about pets and environmental exposures (though we do have an enormous database of research done on laboratory animals), but we can make some educated guesses.

Our pets are likely affected by indoor air contaminants, such as the off-gassing of volatile organic compounds from carpets, curtains, and furniture. Given how much time many of our animals spend indoors, their exposures are likely even more sustained than ours, and some of the nose, throat, and respiratory problems experienced by pets could be linked to air quality. An article published in *Environmental Health Perspectives* actually recommended that pets be used as sentinels of cancer risk from indoor exposures such as radon and tobacco smoke, rather than using laboratory rodents or human epidemiological studies.[1] In other words, pets are like canaries in a coal mine: watching what happens to them after ten or fifteen years of exposure will help us understand what is happening inside our own bodies, since we share the same indoor environment.

In 2008, the Environmental Working Group published a report on the toxic load borne by our pets. They tested for seventy different industrial chemicals in a group of pets and found evidence of contamination for forty-eight of these, including mercury and flame retardants.[2] More recently, consumer watchdog groups have voiced concern about the presence of phthalates in pet toys, particularly the soft vinyl toys that dogs so love to chew. The slobber and the grinding break down the plastic, allowing chemicals to leach into our dogs' mouths. Phthalates may increase risk of liver, kidney, or reproductive problems. Another recent study linked high serum concentrations of PBDEs (flame retardants) to hyperthyroidism in older cats.[3]

A book by veterinarian Michelle Bamberger and pharmacist Robert Oswald called *The Real Cost of Fracking* is one of the first to draw attention to the possible impacts of hydraulic fracturing on our pets. The basic message of the book is that we simply don't know the true costs of fracking, but there is good reason and some data to suggest that exposure of our pets to water contaminated by fracking operations may cause a host of health problems.

Pets are affected by the environment and, like us, contribute (unwittingly, of course) to environmental problems such as global warming, pollution, and unsustainable patterns of consumption. Perhaps one of the largest environmental impacts involved with pet keeping is food. Consumption of meat is one of the key drivers of global warming, so it seems that a world full of carnivorous pets could only make matters worse. Of course, one might argue that since our pets eat a lot of the waste materials from the meat industries (see chap. 17, "Feeding Frenzy"), this isn't a significant impact. But as more and more people are buying human-grade meat for their dogs and cats (a good trend, in my opinion), the meat impact remains significant. I have a suggestion: since our dogs and especially our cats actually need some meat to be healthy, why don't those who acquire a dog or cat or some other meat-eating creature reduce, by equal amounts, their own meat consumption? Then owner and pet cancel each other out, in terms of carbon output.

Pet food production also has an impact on marine ecosystems. Small wild-caught forage fish are one of the key ingredients in fish-based dog and cat foods. Pets, and especially cats, account for 13 percent of the total wild-caught forage fish consumed each year—some 2.5 metric tons.[4] These small fish are vital food for other marine creatures, such as seabirds, marine mammals, and larger fish. Unsustainable fishing harvests are having devastating effects on the oceans, as captured in the documentary *The End of the Line*. In response to concerns about marine ecosystems, some companies are selling cat and dog foods made with fish caught sustainably and certified by the Marine Stewardship Council.

The issue of sustainable fishing raises, pointedly, another ethical concern with pet keeping. In a world in which millions of people are starving or malnourished, does it really make sense for so much of the world's food, and particularly its protein, to be funneled into mouths of pets? This concern is often voiced as pets versus people, but it isn't, and proposals to simply kill off the world's pets—or better, eat them—are off point. (And yes, both of these have been suggested.) More on point would be for breeders to stop increasing the supply of new pets (we already have plenty of supply in our shelters), for industry to stop creating demand, and for consumers to reduce their consumption of pets and pet food. A small population of pets is sustainable; the army we are currently amassing is not.

Pet owners buy food for their animals. They also buy vast quantities of toys, beds, cages, kennels, carrying totes, bowls and bottles, clothing, collars, tags, automatic ball throwers, medicines, and so forth. To give you a sense of the scale of pet keeping, consider this: there are more pets in the United States than there are cell phones—and by a safe margin. And if you think about all the stuff that people buy for their beloved animals, the environmental costs certainly seem to pile up.[5] As with many other consumer products, pets and their paraphernalia are cheap and disposable. (I suspect that the appeal of the paraphernalia is almost as strong as the appeal of the pets themselves.) Like used cell phones, used pet products (and often used pets, too) eventually wind up in landfills.

And we cannot forget the waste materials: fecal output is estimated at about three-quarters of a pound per day for a large dog, which equals approximately 274 pounds of feces a year. This fecal material presents a number of environmental problems, from the large number of plastic poop-pickup bags in landfills, to the potential for contamination of lakes, streams, and rivers with pathogens from feces that is left on the ground. Cat feces has its own set of issues: it must be handled properly to contain potential pathogens, particularly the microbe that causes toxoplasmosis. Thus, we have litter boxes that contain the waste and that can be cleaned regularly. Like baby diapers, though, mounds of litter take up space in landfills. The popular clay-based litter is thought to present health concerns both to cats, who ingest the dust when they dig in the litter box and when they lick the clay off their feet, and to humans, who ingest the dust particles.

It is hard, in a world hurtling toward climate catastrophe, to think that a little bit of dog poop or kitty litter is really going to make much difference. But every small action is important, if nothing more than symbolically. (Isn't living responsibly supposed to be its own reward?) Reducing the number of pets in the world certainly makes sense, since consumption adds up. And since the out-of-control acquisition of pets and pet products is driven much more strongly by industry and advertising than it is by a basic human need for "the bond," it would benefit animals immeasurably if consumers cut back on impulsive purchasing of animals and animal stuff.

A growing number of books about sustainable pet care are helping pet owners sort through both the environmental risks to our pets and the environmental impacts of our pet-keeping practices. I particularly like Carol Frischmann's *Pets and the Planet*. She has chapters on how pets affect the environment, why choice of pet matters, how to shop for Green pet food, equipment, and toys, how to deal with pet waste, how to manage risks from household exposures, and how to find eco-friendly pet services, such as grooming, boarding, and veterinary care.

The considerations thus far in this chapter have been limited to "normal" pets—the dogs, cats, rabbits, hamsters, and so forth of the world. But there are serious environmental problems with the so-called exotic pets, which are increasingly popular both in the United States and elsewhere. Demand for "weird" pets is high and is encouraged (irresponsibly, in my opinion) by the media and pet industry. For example, *Animal Planet* regularly airs shows featuring "top 10 strangest pets," almost always focused on the bizarreness of the animal and not welfare considerations of acquiring and keeping a kiwi bird, say, or a capybara. The so-called Hollywood effect—the sudden interest in certain species after the release of a film with an animal star—is well-documented in relation to the purchase of certain dog breeds. *101 Dalmatians* caused an uptick in Dalmatian sales, and *Beverly Hills Chihuahua* did the same for those little bundles of nervous energy. Sudden demand leads to an increase in breeding and pricing and, then, to the inevitable downward curve when the shelters fill up with unwanted Dalmatians and Chihuahuas (both of which are considered "high maintenance" breeds and not appropriate for the faint of heart or impulse consumer).[6] This same phenomenon occurs after "wildlife" films, too. *Finding Nemo* created a surge in demand for the clown fish, and *Rio* caused a sudden demand for blue macaws. Both species are captured from the wild and sold (if they survive that long) to brokers and individual pet owners.

I'll talk about welfare concerns for exotic animals in later chapters; here I'd simply like to note that the international trade in exotic pets is one of the key drivers of biodiversity loss. "Exotic" is defined differently in different contexts, but here can be taken to mean any species of animal that has not been domesticated. Many of the birds, reptiles, and mammals destined to become someone's exotic pet are wild caught or, in the case of birds and reptiles, taken from the wild as eggs. As researchers from the Wildlife Conservation Research Unit at University of Oxford write, "Unsustainable harvest of wild animals for the pet trade has already led to population decline and collapse for many species."[7] The radiated tortoise, for example, is critically endangered and will likely become extinct. But you can still buy one over the Internet. This is a global problem, as exotic pets are sourced from all over the world, destined for pet owners all over the world. Demand for exotic pets is especially high, at the moment, in the Middle East and Southeast Asia. The global demand for pets has kept pace with, and even in many places exceeded, human population growth, and this does not bode well for conservation efforts.

Some of the movement of animals is legal; much of it is illegal. It is illegal, for instance, to transport endangered species across borders. But it happens all the time anyway. Traffickers stuff birds into their socks, line their shirts with frogs, and use all manner of other imaginative methods to sneak animals aboard airplanes bound for pet-hungry lands. Illegal trade in exotic animals is often linked with drug and weapons trafficking and with terrorism, since capturing and selling wild animals or their body parts is a relatively easy way to make money.[8] So purchasing exotic pets ties a buyer into a whole invisible network of violence.

As you can see, local decisions really do have global reach.

Worrying about Spot

23. Turn Me Loose

The novel *Room*, by Emma Donaghue, tells the story of Jack, a five-year-old boy who lives in a single room with his Ma. Because his entire life has been spent in Room, this is his reality—he doesn't know that there is a world outside Room. Room is perfectly humane: Jack and his Ma have food provided daily by their caretaker (captor) Old Nick, who also comes in nightly to sleep next to Ma. Room is clean, they have clothes to wear, and a place where private business can be accomplished in a sanitary fashion. And yet, as you will likely have suspected, the novel is a horror story. It is spine chilling, even though there isn't any blood or guts. It reminded me, when I read it, of what we do to our pets. In particular, it reminded me of what I have done to countless rats, geckos, snakes, hermit crabs, hamsters, and guinea pigs. I have put them in Room.

People don't necessarily think of pets, especially dogs and cats, as captive animals. But they are. And for many pet animals, captivity involves being kept in a cage for their entire lifetime (the necessity of bars being a good indication that we hold these animals against their will). Sometimes captivity seems well-balanced against a measure of freedom, such as a dog who is kept inside a house but who shares his daily life experiences with a special human or human family, has plenty of opportunity to get outside, run free, follow her nose. Other times, and perhaps for the majority of creatures, captivity involves constant confinement. Many pets spend their entire lives alone in a tiny cage or tank. This isn't a minor animal welfare issue we're talking about. An estimated 80 million dogs, 95 million cats, 160 million fish, 20 million birds, and millions of diverse other animals are kept as pets in U.S. households.[1]

Life for permanently cage-bound creatures is rather like life in prison and even, in some cases, like solitary confinement. These animals have no meaningful interactions with their own kind, have minimal physical activity, have almost no mental stimulation, and experience a severe reduction in the kind and quality of environmental stimuli. Like humans, our pets suffer psychological and physiological ill effects from these conditions. As research into animal cognition and emotion further expands our understanding of animals' inner lives, the more ethically problematic holding them captive becomes.

Things are particularly bad for captive reptiles and amphibians. Although I didn't recognize it as such at the time, I am certain our gecko Lizzy was slowly going crazy in her twenty-gallon glass tank. After we had had her for several months, she began to claw repetitively at the side of her tank, as if she were trying to go for a walk and couldn't see that there was glass in the way. Clifford Warwick, who has written about the morality of keeping reptiles as pets, describes "at least 30 captivity stress-related behaviors . . . regularly observable in most kept reptiles . . . such as hyperactivity and interaction with transparent boundaries, both of which involve persistent attempts to escape, and hypoactivity, which involves efforts to biologically 'shut down' from a poor environment." Many of these signs, says Warwick, are "simply ignored by keepers as irrelevant."[2] I got a pit in my stomach when I read this.

Reptiles have a level of sentience and sensitivity "comparable with other animals, including humans," according to Warwick, and new research is continually adding to our understanding.[3] For example, a study by Anna Wilkinson and colleagues published in *Animal Cognition* in 2014 provides evidence that the bearded dragon (*Pogona vitticeps*) is capable of social learning through imitation, a skill previously thought to be limited to "higher species" like humans and certain primates.[4] Yet reptiles are different from dogs and cats in important respects. Reptiles are not and have never been domesticated, nor is it likely that they ever could be. They are too different from us. Unlike dogs and cats, who can have "life-sharing associations" with humans (they can share the same physical and emotional space), reptiles cannot share our life space and are typically caged for their entire lives. "This," says Warwick, "takes evolved life management—survival programming—out of the control of the reptiles." Reptiles belong to a different biological class and typically come from parts of the world with very different ecosystems. We know little about the biological needs of reptiles, "and what we do know appears to indicate that we do not, and probably cannot, provide for them."[5] And, indeed, reptiles in captivity almost never do well, and mortality rates for pet reptiles are shockingly high. At least 70 percent of captive reptiles will die before they even reach the pet-store shelf, and of these survivors, about 75 percent will not live past their first year as a pet.[6] Lizzy lived for a little over two years, out of a natural life span of about twenty-five. Warwick asserts that reptiles simply should not be kept as pets, and I find myself in agreement.

Fish are no better off. Fish tanks are often barren, particularly for the goldfish sold by the billions as easy-to-care-for children's pets. Goldfish are billed as small, low-maintenance, and short-lived, but that's only because their growth is stunted and untimely death is inevitable. Goldfish are not naturally small and will, in ideal conditions, grow up to a foot long and live up to twenty-five years. They are also—contrary to what many people assume—quite intelligent and easily bored. The prototypical goldfish in a small bowl on the dresser is, in fact, a travesty. If the pet industry were taking animal welfare seriously, goldfish wouldn't be "starter pets" and tiny tanks wouldn't even be on the shelves. Yet, instead, tanks seem to be shrinking.

One of the trends touted in pet industry trade journals is the so-called nano tank (also referred to as the micro or pico tank). These are fish tanks that appeal to the customer who wants something very compact, perhaps to fit unobtrusively on a desk or kitchen counter. "People have always loved little bowls . . . people love that an eye-catching aquarium can exist in such a tiny framework and they don't have to maintain a huge setup."[7] And small they are. The Contour Aquarium comes in three- and five-gallon sizes; the BioBubble is a compact three and a half gallons; and the Cue Desktop is a mere two and a half gallons. The USB Desk Organizer Aquarium—with attached desk organizer, including a multifunction pen holder and LCD calendar—holds one and a half quarts of water (that's six cups) and is advertised as "suitable for live fish." If you are selling nano tanks to customers, says *Pet Age* magazine, you should definitely also be selling the "tiny livestock" to put in them. Some of the suggested species include micro rasboras, danios, tetras, killifish, and shrimp. In what kind of sick and twisted world is six cups of water considered an ideal life habitat for an intelligent fellow creature?

I found another iteration of the nano tank in *Pet Business* journal: a full-page advertisement for Betta Falls. The plastic tank is has a three-tiered curved design, and each tier is meant to house one betta. Each of the three tiers is separated by a frosted panel, so the fish cannot see each other. The 16 × 11 × 11–inch tank is compact enough to fit on your desk, and requires only two gallons of water. This means that each fish has about ten cups of water in which to live out his entire life. Unfortunately, Betta Falls gets bad reviews from customers, who say that the pressure from the waterfalls is so strong that the fish has to stay off to the side, decreasing his already tiny swimming space. Also, the fish get sucked into the filters, and wind up dead. Pet stores often try to convince customers that bettas like to live in small cups of water by spinning a fascinating tale of natural history: in the wild, they tell us, bettas actually live in the watery footprints of water buffalo. But this isn't exactly true. The natural habitat of *Betta splendens* is the slow-moving and thickly vegetated river basins and rice paddies of the Mekong and Chao Phraya Rivers. During the dry season, when much of the river water evaporates, fish can get trapped in small water pockets and, yes, even water buffalo footprints. The fish have evolved several adaptations to survive entrapment in puddles: they are very good jumpers, and they have a highly vascularized gill structure that allows them to supplement their oxygen intake from the surface (stagnant water often lacks oxygen). This is why bettas can survive longer than most fish would in little tiny plastic cups on a shelf in Walmart, and also why the cups have lids.

Like pet reptiles, amphibians, and fish, small mammals often spend their whole lives confined to a tiny enclosure. For example, the so-called pocket pets—rats, mice, hamsters, gerbils, guinea pigs—are typically housed alone, in a small enclosure made of plastic or metal or glass. Many of the cages sold in pet stores are far smaller than the cages used in laboratory or research settings, and in fact would fail welfare standards established to protect research animals. There are no welfare standards regulating cage size for pets. These creatures never feel the earth under their paws, only the rough wood shavings that are sold as bedding in pet stores. They eat the same bland "pellet" every day. They never get to forage or otherwise work to obtain their food, even though this is a deprivation: laboratory studies have shown that many species of animal will choose to work for food even when offered the choice to get it for free.

Small mammals kept as pets often have no same-species companionship, despite the fact that most of the common species of pocket pet are, in nature, quite sociable and have complex interrelationships with others of their kind. Their only time out of the cage might be when they are handled by their captor. Their Room is their entire reality. (I wonder if they have an evolutionary memory of their natural habitat and, thus, a deep-down sense of what they are missing.) I've never seen research on the mortality rates of pocket pets in the home, but I suspect that the numbers are quite high, though perhaps not quite as high as for reptiles since mammals are warm-blooded and we can more easily mimic at least some aspects of their physical environment. Still, life in solitary confinement is not an ethically sound option for these creatures and, like reptiles, they probably cannot be kept as pets without causing them suffering.

The keeping of small animals highlights one of the most significant harms we impose: social isolation. The more social a species, the more significant the harm of solitary confinement. We may think of ourselves as our pet hamster's best friend, but he does not think of us in the same way. Hamsters are not programmed, genetically, to bond with humans. They may become tame and view us without a great deal of fear, but we are not of their kind. Even in our company, they are alone. In our tendency to be individualistic, we may fail to recognize how crucial sociality is to welfare. New research is suggesting that social isolation may cause not only emotional suffering but actual physical harm as well. A study of gray parrots showed that social isolation actually shortened telomeres and, thus, life span.[8] Parrots who were housed alone died younger than those who had a parrot companion. Which common pet species are social? Nearly all of them. Dogs and cats, a vast majority of the fish species, rats, hamsters, rabbits, gerbils, hermit crabs, bearded dragons. Probably a better question to ask is which animals are solitary in their habits. Yet even relatively "solitary" animals like the leopard gecko interact with others in their natural environment and cannot be said to be truly solitary. There is no such thing as a solitary animal.

So far, I've talked about confinement as isolating, like the singleton hamster living in his own private Habitrail. There is another side to confinement that can also be problematic for animals: overcrowding. If you've ever been on a busy subway car or on a jam-packed airline flight, you will sympathize with the feelings of anxiety and bubbling rage that can percolate when too many social animals are crammed into too small a space. The upside is that the animals have company; the downside is that they are too close, are forced to interact too often, and are more likely to have aggressive interactions—all of which are stressful. Overcrowding is very common in pet stores and animal wholesale warehouses, where dozens of animals may be packed into a small cage or Tupperware, like sardines in a tin can. It is an odd fact that a hamster or rat or fish might spend the first few weeks of life in an unbearably crowded situation and then spend the remainder of his life in total isolation.

One of the most common overcrowding situations is the fish

tank. A betta in a nano tank or a singleton goldfish in a tiny bowl may seem cruel; it is equally hard for fish to be overcrowded. At our local pet store, the fish are kept in tanks so full that they appear to be inhabited by one single pulsating organism. Ronald Oldfield, a biology lecturer at Case Western Reserve University, studied the effects of tank size on aggression in a common aquarium fish, the Midas cichlid. He found that cichlids kept in aquariums of sizes generally sold in pet stores to fish hobbyists behaved more aggressively toward each other than cichlids housed in larger, more complex habitats. The increased aggression by dominant fish has welfare implications for the other fish in the tank, who are stressed out by being bullied. In addition to having physical damage, the bullied fish engage in more cowering behavior, may not eat as much, and have a decreased resistance to disease. Dr. Oldfield told the *New York Times*, "If people kept dogs in these conditions, they'd be put in prison."[9]

In considering the ethics of captivity, dogs and cats each present a special case. Both of these species can and usually do live outside a cage. They may be confined to a home some of the time, perhaps even all of the time, but we can provide lives for them in which they have considerable freedom of movement, can display many of their natural behaviors, and (under good conditions) have physically and psychologically stimulating experiences. Both of these species are domesticated, both have coevolved with humans to some extent (particularly dogs), and both can form intimate and meaningful relationships with us.

Psychology professor and dog cognition researcher Alexandra Horowitz argues that what freedom and captivity mean to dogs is unique and must be considered in the context of the human-canine dyad.[10] There is no such thing as a "wild" dog: the entire species is captive. (Feral dogs aren't really wild animals because they still cohabitate with humans.) We control everything about their lives, and they have evolved to be dependent on us. Domestication, she suggests, has changed the nature of dogs, chiefly by "suppressing the pre-domesticate's *Merkelt*, or perceptual world."[11] Sensory sensitivity has been selected against, and dogs have been diminished cognitively. But in other ways, they have been enhanced, such as in their capacity to manipulate the environment using humans as tools. "The very process of artificial selection holds dogs captive, tethered to persons in body and mind."[12] Dogs are inextricably bound to their domesticators and are "naturally captive." (This is my phrasing, not Horowitz's.)

Even though dogs as a species are naturally captive, there is still an important sense in which individual dogs can be and want to be free: They want to be dogs. Yet typical dog ownership involves numerous interventions to circumscribe "normal" dog behaviors such mating, marking, barking, and roaming. We build fences, attach collars and leashes, have dogs neutered and spayed. Dog training involves a careful and methodical process of constraining natural behaviors. The expression of "normal" behavior is routinely violated, and, indeed, violation is integral to pet-keeping practices.

These restrictions on normal behavior, says Horowitz, can have a negative impact on dogs by winnowing down their experiential window into the world. The restriction placed on dogs by, let's say, being always tethered to their human owners is a serious one. "Subjected to a person's decisions about everything from where to walk (down what routes, and when), whom to approach (which dogs and people), and what to investigate (which odors can be loitered on and which cannot), the dog has little independent choice."[13] There are many ways in which we can enhance the freedoms of dogs, even within the broader context of species-level captivity. The "least captive dog," for example, is free from the leash, is not restricted to a house, is able to smell what he wants and when, and can approach who he likes and avoid who he doesn't.

Cats and captivity present, in my view, one of the most vexing pet-keeping conundrums.[14] The way we live with cats has changed over time perhaps more than any other form of pet keeping, even dog ownership. Cats are no longer semifree, as they were for thousands of years. The vast majority of cats are desexed, and because of cat litter and cat food, many more cats live indoors, either full time or part time. Cats are often confined to even smaller spaces than dogs, based on the faulty assumption that cats don't need space and don't need exercise. Indeed, the new orthodoxy of responsible cat ownership, espoused by the Humane Society of the United States (HSUS), is that cats *must* be kept indoors. According to HSUS, this is for the cats' own good: it is dangerous for them outside. But many cat owners, myself included, feel that cats lose something important when denied access to the outside world and that the risks posed by the outside world are ones that at least some cats would happily trade for freedom.

There is an epidemic of obesity among cats, as well as an epidemic of behavioral problems that may stem from boredom and frustration and that result in large numbers of cats being relinquished to shelters and killed. Something about the way we keep cats these days is not working, and I would place my bets on the source being their increasingly restricted lives. Nearly the entire range of their normal behaviors have been truncated.

Some scholars have suggested that as our relationships with animals become less functional, and the more that the animals are integrated into human families, the greater their risk for poor welfare. This seems counterintuitive at first. Wouldn't a family dog be treated better than a working dog? Wouldn't an indoor cat be more pampered than a barn cat? But maybe pampered is not what our animals really want. The way we live with animals now makes it more difficult for them to engage in normal behaviors. In addition, cats and dogs are exposed to an increasing number of stressors like boredom and loneliness. Increased urbanization leads to smaller living spaces. More dogs and cats live in big cities. People travel often and move frequently—dislocations that may seem okay to us but can be extremely stressful for some animals. Even the increased affluence of pet owners might not be an unalloyed good for our cats and dogs. For example, one study suggests that although affluence can mean more veterinary care and a premium diet, it often also correlates with less attention to social and behavioral needs and less exercise. Decreased levels of interaction with an owner can manifest in separation anxiety or obsessive compulsive disorder, for both dogs and cats. And lack of mental stimulation can decrease behavioral flexibility and promote reactivity and depression.[15]

When it comes to the harms and benefits of living in captivity as pets, each species and each individual animal needs to be considered on their own. Some animals, like dogs, can flourish and can live with a high degree of freedom. With other animals, the best we can hope for is controlled deprivation. It is hard to make blanket statements about captivity, except to say that it is morally significant and needs to be on the top of our list of welfare concerns for pets.

24. A Boredom Epidemic

Nearly all pet animals suffer from boredom at one time or another, some persistently and pathologically. As caretakers, we may be totally unaware of animal boredom, or perhaps we are aware but aren't sure how to combat it. The most serious cases are usually found among animals who are cage bound—rats, birds, guinea pigs, goldfish, hermit crabs—but other pets such as dogs and cats can also be affected.

Research in cognitive ethology has by now firmly established what is probably already obvious to you: animals have feelings, have an interest in their own lives, and do things because they enjoy them. Beyond this, we know that an animal who is well-fed and safe from predators can still suffer from having nothing to do. Biologist Françoise Wemelsfelder provides one of the most comprehensive explorations of animal boredom. As Wemelsfelder explains, for human and nonhuman animal alike, activity and sensation are not in and of themselves meaningful when disconnected from voluntary, authentic interest.[1] To be interesting, activities and sensation must engage an animal in "attentional flow." Nonhuman animals experience attentional flow when they organize and engage in explorative activities or spontaneous play, alone or with other animals. "Through attentiveness, exploration, and play, animals engage with the environment for the sake of interaction in its own right."[2] The form and style in which attentional flow occurs "may vary throughout the animal kingdom; however, expressions of voluntary attention have been observed right down the phylogenetic scale," and we shouldn't exclude any species from this approach. "I am of the opinion," she writes, "that voluntary attention reflects a general principle of behavioral organization."[3]

Caged animals, she goes on to say, have "very few opportunities . . . to express individual interests or preferences. The cage environment may contain noise, smells, and things to see, and to these, the animal may respond; but this is not to say that the animals can engage creatively and experience attentional flow."[4] The severe restrictions of horizontal and vertical space that cages impose don't allow animals to move more than a few steps in any direction and severely restrict their field of view. One primary effect of this restriction: animals cease to be active or attentive. They may "go through extended periods of motionless sitting or standing, often with drooping heads and ears, half-closed eyes, abnormally bent limbs, and pressing themselves against a wall."[5] Or they may spend most of their time lying down or sleeping. Pet owners could mistakenly interpret sleeping to mean that their little creature is happy and unafraid of his surroundings, while this is not at all what's going on.

As their confinement drags on, animals develop abnormal behaviors, which often progress into stereotypies, that is, repetitive behaviors such as pacing, circling, or rocking. These stereotyped behaviors emerge when an animal cannot engage in behavior it is highly motivated to perform, such as hunting for food, seeking social interaction, or trying to hide from predators. An animal in this situation enters into a vortex of psychological and physiological degeneration from which it is hard to reemerge. She gradually loses her behavioral versatility and open-endedness. The signs of chronic boredom appear to be the same in animals as they are in humans: apathy, listlessness, compulsive habits, frustration, restlessness, hostility, and the disappearance of inquisitive play.[6]

25. Don't You Want Me?

In March of 2013, thirty-seven-year-old Hojin Lee was arrested on animal cruelty charges. When Lee was evicted from his Park Ridge, Illinois, apartment, he left his terrier mix named Bruno behind, like a box of unwanted belongings. Bruno was finally discovered by an apartment manager after a month and a half without food or water, severely malnourished and nearly dead. In an act of brazen meanness, or simple inattention, Lee had left a bag of dog food on the counter, just out of reach for the small dog.[1] This case may sound extreme, but such overt and cruel forms of abandonment are surprisingly common.

Animals are also tossed from car windows, left alone on country roads, deposited during the wee hours of the night into the shelter's overnight drop box, specially designed for people too cowardly to abandon an animal in broad daylight. People move, or get married, or get divorced, or decide to go back to school, or decide to go on an extended vacation, and simply leave their animal behind. (What are shelters for, after all?)

We abandon our companion animals in small ways every day, too, by failing to provide consistent emotional connection when we are home or by leaving them alone for long hours while we go about our own business. We foster relationships of dependence, where we are the sole source of food, exercise, and social companionship. And then we, who remain independent, leave.

The tragedy of abandonment is perhaps nowhere more poignant than in the case of dogs. One of the key goals of the dog owner is to encourage the dog to form a strong social attachment. And attach they do. It is what we love about dogs—their unwavering loyalty to us and only us. The reason why breeders, trainers, and self-proclaimed dog experts consider it ideal to buy a puppy at eight weeks of age is to take advantage of the socialization window, so that the young pup attaches to her human instead of to her own mother and siblings. Secure attachment can benefit dogs by making them feel happy and safe in our company; it can also cause them stress (e.g., when the attachment object is gone for long periods) and heartbreak. We want them to be dependent on us. But when dependence causes behavioral problems such as separation anxiety, the dog often takes the blame.[2]

If abandonment takes place after an affectional bond has formed between dog and owner, it will be particularly emotionally distressing for the dog. For instance, dogs entering a shelter after being abandoned by their attachment object often "experience a sudden and strongly traumatic bond disruption."[3] It also appears that dogs will readily forgive—another thing we love about them. In one shelter study, even very brief positive interactions with humans (three ten-minute periods) were enough to again promote attachment behavior in an abandoned dog.[4]

26. Cruelty, Abuse, Neglect

The general pet-owning public may underestimate how prevalent animal abuse is. In some ways, the media exacerbates the problem, since most of the things we see and read about pets are warm and fuzzy: 90 percent of people talk to their pets; 66 percent tell their pets "I love you" every day; more than half share their beds, take their pets on vacation, and celebrate their pets' birthdays. We can think to ourselves, "Oh, isn't the human love affair with dogs and cats precious?" But the people who are captured in these industry surveys are a certain kind of pet owner. These are precisely those who typically *do* love their animals and make them part of the family. The uneven focus on the good sides of pet ownership helps us ignore the fact that a great many pet owners are irresponsible, neglectful, and even cruel. And, it also helps us look right past the fact that some people who tell their dog "I love you" every day and dress him up in fancy collars might be the same people who leave the dog alone all day, with far from enough physical and social stimulation.

The most obvious cost to our pets of our pet-keeping practices is the abuse they suffer at the hands of mean-spirited or emotionally disturbed people. The scope and lengthy catalog of human cruelty to animals will make you wonder whether the price animals pay for living in our midst isn't offensively high. Prosecutable forms of animal cruelty are just the tip of the abuse iceberg: animals also suffer less blatant but more widespread forms of physical and emotional abuse such as punitive training methods, prolonged confinement, boredom, and exposure to incessant teasing. These are not rare and unusual occurrences. My impression is that more companion animals are mistreated than not and that well-cared for and beloved pets are in the minority. You may scoff at this, but I invite you to read the statistics, the forensic textbooks, and the daily news and see for yourself.

On any given day, a large handful of heartbreaking and horrific stories of animal abuse will be made public. On the day I write this, for example, I read a story about an eight-month-old collie mix picked up by animal control. The puppy, who rescuers named Lad, had been shot in the face. The gunshot wound was several days old by the time Lad was picked up. He was suffering from serious infection and from starvation, since his shattered jaw kept him from being able to eat or drink.[1] In the same week, a Florida man punched and kicked his dog in an apparent attempt to get the dog to bite a bystander.[2] A New Jersey man was arrested after dragging his dog by a leash behind his car (he claims that he forgot the dog was tied up).[3] Another Florida man attacked and killed his dog Nelly with a pickax to the head because she "looked at him funny."[4]

Here are some chapter titles from Leslie Sinclair, Melinda Merck, and Randall Lockwood's *Forensic Investigation of Animal Cruelty*: "Thermal Injuries," "Blunt Force Trauma," "Sharp Force Injuries," "Projectile Injuries," "Asphyxia," "Drowning," "Poisoning," "Neglect," "Animal Hoarding," "Animal Sexual Assault," "Occult and Ritualistic Abuse," "Dogfighting and Cockfighting." Humans are imaginative when it comes to forms of animal abuse. Run, Spot, run!

Sinclair, Merck, and Lockwood define cruelty broadly as "any act that, by intention or by neglect, causes an animal 'unnecessary' pain or suffering."[5] But there is clearly a difference between acts that all of us would agree are cruel (smashing a kitten with a shovel) and those that are more personal and subjective (leaving a dog in the backyard all day, every day).

There are no consistent legal definitions of abuse, cruelty, or neglect, and laws to protect animals are uneven and weak and vary from one state to another. Anticruelty laws are often vaguely worded, leaving prosecutors and judges unsure how to interpret them. For example, the animal cruelty statute in Washington tells us that "a person is guilty of animal cruelty in the first degree when, except as authorized in law, he or she intentionally (a) inflicts substantial pain on, (b) causes physical injury to, or (c) kills an animal by a means causing undue suffering."[6] A man accused of blowing up his daughter's yellow Lab could not be accused of cruelty because, as the undersheriff explained, "the dog died instantaneously and didn't suffer before it died."[7] Who is responsible for pursuit and prosecution of crimes against animals? Again, there no clear answer—it varies from place to place.

It is hard to know precisely how many animals are abused each year or by whom since there is no federal database of statistics related to animal-cruelty crimes.[8] The data on animal abuse has been collected, piecemeal, by various animal advocacy groups over the past several decades. The Humane Society of the United States estimates that there are hundreds of thousands of animal cruelty cases a year. This is probably a conservative estimate. The primary groups of abusers appear to be domestic violence batterers (heterosexual and same sex); people with personality disorder; children and teens, primarily males and primarily preschool age; hoarders (75 percent of whom are women); adult women and men with unrealistic expectations about how an animal should behave; and violent offenders (rapists). Dogs are the most frequent target of abuse, with cats a close second.[9]

In September of 2014, the director of the Federal Bureau of Investigation announced that cruelty to animals will now, for the first time, be included in the Uniform Crime Report Program. It will no longer simply be grouped into the all-other-offenses category of minor crimes but will be classified as a distinct group of offenses, including simple and gross neglect, intentional cruelty or torture, organized abuse (such as dogfighting), and sexual abuse. Having a national database of incidents and arrests is an important step in efforts to combat cruelty because it will help advocacy groups and law enforcement better understand, prosecute, and seek to prevent crimes against animals.

Given how strongly people advocate the benefits of owning a pet for children, it is unsettling to know that a great deal of animal abuse is meted out by children. Pamela Carlisle-Frank and Rom Flanagan write, "As researchers have discovered, evidence of animal abuse in childhood is alarmingly high."[10] In several different surveys of college students, about half reported having either observed or participated in animal cruelty when they were young. "Out of these, 20% of respondents had actually abused animals as children, one in seven had killed a stray animal . . . and 3.2 had killed their own pets."[11] Males appear more likely than females to participate or observe.

Nobody really knows from which dark places of the soul an individual's impulse toward cruelty arises. But in trying to understand cruelty toward animals, researchers have identified certain motivations, among them the desire to control an animal or to retaliate against an animal for some perceived insult, prejudice against a species or breed, undifferentiated aggression, a desire to impress or shock other people or to exact revenge against a person, and plain old sadism.[12]

People who abuse animals often feel that their abuse—which they often label "good discipline"—is justified by the behavior of their pet. A pharmacist friend of mine recently told me how one of her coworkers reported, with a certain amount of pride, that her fiancé had punched his dog in the face as punishment for running away. Ironically, these disciplinarians are often the same people who fail to learn anything about effective and appropriate training techniques, which behaviorists almost unanimously agree should be based on positive reinforcement rather than punishment. In a vicious cycle, the harsh punitive approach can make a dog or other animal extremely stressed out, which in turn may manifest in more behavioral issues. Journalist Mark Derr writes, "To my mind, any theory of learning or behavior modification that relies on the asymmetrical exercise of power in such a way that it inflicts pain or suffering on a weaker being is wrong." Aversive training "retards learning in the long run" and can "give rise to violent aggressive behavior."[13]

Those who abuse animals also tend, at least according to research, to be more sensitive to stressors, and abuse may be one coping mechanism. The pet's own behavior is often a source of stress, which then precipitates abuse, which then likely makes the animal even more anxious, leading to an escalation in behavioral issues. There are vicious cycles everywhere in this realm of pet keeping. Pets can, of course, be extremely frustrating and stress inducing—something people don't always consider when in the contemplating-getting-a-pet stage. When tension levels rise, people may have a lower threshold for handling, with patience, the daily challenges of living with an animal. I experienced some of this myself when our Colorado town was struck by severe flooding in the fall of 2013. We were evacuated from our home and large sections of our town were completely destroyed, including the infrastructure that provided water, sewage, gas, and electricity. The most challenging aspect of the entire two-month ordeal was dealing with my pets, particularly my two dogs. Small annoying habits such as Maya pulling on the leash and Bella's ear-piercing bark became magnified almost to the point of being intolerable because my stress level was so high. At one point, on a particularly hard day, both dogs ran off and jumped into a drainage filled with *E. coli*-infested flood water—after I had made a huge effort to take them to one of the few places they could be off leash—and I found myself in tears, yelling at my poor wet dogs, "You are getting on my last *nerve*! One more stunt and you are going straight to the shelter!" It gave me some perspective on why so many animals did actually wind up at the local shelter after the flood, and why quite a number were never reclaimed by their owners. When you are up to your armpits in mud, pets may be more than you can handle.

27. A Hidden World of Hurt

In discussions of animal abuse, most of the attention goes to physical abuse. Yet emotional abuse and mistreatment of animals is every bit as important as a source of widespread suffering. (I think, with shame, of my flood-induced tirade against Maya and Bella, and of them standing with their ears pulled back and tails tucked under.) As of this writing, none of the fifty states has animal cruelty statutes that include language specifically acknowledging emotional neglect, abuse, or suffering. Some even explicitly say that injury must be physical in nature. This is a huge deficiency, especially in light of the past three decades of research into animal cognition and emotion. As with children, physical abuse of animals also invariably involves psychological suffering (from fear, anxiety, and so forth), and we ought to consider an animal who is physically maltreated to also be emotionally maltreated.[1]

Emotional cruelty is a serious problem for pets. But emotional neglect is perhaps the more insidious and widespread concern because it largely flies under the radar of awareness. Frank McMillan argues that emotional neglect occurs when a caretaker fails to provide an environment in which an animal has social companionship, mental stimulation, a sense of control, a sense of safety, protection from danger (e.g., hiding places), and adequate predictability and stability to life events.[2] Of course, trying to determine exactly what level of social companionship or mental stimulation or sense of control a particular animal needs is a challenging task, open to significantly more subjective judgment than, say, whether an animal has been physically injured or deprived of food and water. Do we take into account the intent of the human? Or just the harm felt by the animal? When does "normal" treatment—such as confinement in a crate—move from being acceptable into being abusive? After one hour? Twelve hours? Three days? Three months?

Neglect can stem from ignorance, lack of motivation, poor judgment, and high levels of frustration with an animal. Although poor judgment and lack of motivation are hard to fix, ignorance and frustration can often be addressed through education about how to care for an animal and how to interpret and shape an animal's behavior. Child-protection workers have been able to educate parents on how to create nurturing, stable homes, and so, too, might animal-protection workers be able to educate pet owners.

It is imperative that cultural and legal standards for the care and treatment of companion animals take into account their emotional needs. There is a great deal of work to be done in carefully detailing the forms and manifestations of emotional mistreatment of companion animals and then drafting legislation that will protect animals from these kinds of harm.

28. Quiz: Cruel Practices

Please mark the following as *A* (cruel or abusive) or *B* (acceptable):

1. Tethering of a dog in backyard
2. Single hamster kept in cage for lifetime
3. Goldfish in small bowl
4. Denial of food or water
5. Denial of veterinary care
6. Failure to brush a dog or cat's teeth once a week
7. Denial of affection
8. Unclean living conditions
9. Declawing cats
10. Outdoor cats
11. Indoor cats
12. Bird in a cage
13. Pinioning a bird's wings
14. Shock collar on dog
15. Electric fence for dog
16. Obesity in a dog or cat
17. Tail docking
18. Ear cropping
19. Debarking
20. Failure to provide adequate pain medicine to an ill or injured animal
21. Screaming at your dogs after they jump in *E. coli*–infested water

29. The Strange World of Animal Hoarding

John was a dear, sweet man. He lived alone for most of his life, in a small apartment in Anaheim, California, struggling to make a living cleaning houses while trying to get work as an actor. He was openly gay during a time when homosexuality was feared and misunderstood. He read Shakespeare and had a close network of friends in the gay community. He also loved dogs and tried to save as many as he could from homelessness and euthanasia. Near the end of his life, he had at least ten and perhaps as many as fifteen dogs living in his tiny home, and it had become a mess of feces and hair and dirt. His friends fretted over his living conditions but couldn't make him see the problem. He was a hoarder.

Animal hoarding is perhaps our most enigmatic form of abuse. Hoarders are pathological animal collectors, often "rescuing" multiple animals, usually dogs or cats, from shelters. They profess their love for animals yet fail to adequately care for them. Hoarding doesn't quite fit into the category of deliberate animal abuse, nor does it really fit into the "neglect from indifference" category. Indeed, hoarding is an unusual form of abuse because it is often the result of too much empathy or concern for animals. It may initially begin as an effort to save unwanted animals, particularly from euthanasia. However, as the animals pile up, the hoarder becomes less and less able to care for them all.

The conditions in which hoarded animals are kept are often horrific. “Words, and even pictures,” says veterinary epidemiologist Gary Patronek, “are insufficient to describe the extent of squalor and suffering endured by animal victims of hoarding . . . extensive accumulations of feces and urine throughout the house, sometimes many inches deep, resulting in structural damage so extensive that demolition is necessary.” The air inside a hoarder’s house becomes toxic from the accumulation of ammonia and can result in “stench so pervasive that exposed clothing cannot be adequately washed and must be discarded.”[1] Dead animals are often left where they died or are systematically stored in some fashion (e.g., sorted by fur color). Often there is no water or only dirty water, and no food or spoiled food. Death from starvation is common, and animals often have to fight for food and can be forced into cannibalism. They have little space to move around and no opportunity for exercise. They have untreated illnesses and injuries and parasites.

The number of animals kept by a hoarder ranges from as few as ten to several hundred. The popular stereotype of a hoarder—single female, living alone, not working outside the home—is largely supported by the data but is not exhaustive. Hoarders are also men, married couples, young, and old. Hoarders typically satisfy criteria for adult self-neglect and have behavioral deficits such as an emotional attachment to and need for control over possessions (animate or inanimate) and lack of insight into their own behavior.[2] Patronek estimates that there are at least five thousand cases every year, with up to a quarter of a million animals subjected to this kind of abuse.

Mental health professionals don’t fully understand why people become hoarders. One possible explanation offered by Patronek is that childhood experiences with a pet may lay the groundwork. For children in dysfunctional families, a consistently receptive pet may become a means of escape and a substitute for human relationships.[3] This emotional dependence on animals may evolve into hoarding behavior. Whatever the source of this mysterious behavior, hoarding results in profound suffering for animals.

30. The Links

Many of the early voices in the animal protection movement were also speaking out against child abuse and were involved in the abolitionist movement. This is not a coincidence. Research over the past several decades has confirmed what the early pioneers of the humane movement intuited: the abuse of animals is frequently linked with violence toward humans, including the mistreatment of children, as well as intimate partner violence and elder abuse. Animal abuse is also connected with antisocial behavior, criminality, and mental disorders. These various connections are referred to as "The Link."

Most people have been exposed to the idea of The Link, through the popular claim that serial killers and school shooters spent their childhoods doing unspeakable things to animals.[1] Research actually backs up this claim and goes well beyond it, too. Animal cruelty in childhood is a fairly strong predictor of violence toward humans in adulthood and co-occurs with other types of antisocial behaviors and personality traits.[2] Those who engage in animal cruelty are significantly more likely to be violent offenders than nonviolent offenders, and animal cruelty is a better predictor of sexual assault than previous convictions for homicide, arson, or firearms violations. Animal cruelty is also associated with substance abuse.[3]

Perhaps even more significant than the associations between animal abuse and criminal behavior are the links among various forms of domestic violence. Animals and children are often abused in the same household, as are animals and spouses. Moreover, animals often become a tool of violence: the animal is harmed in order to hurt, scare, or intimidate. In one study, almost three-fourths of pet-owning women seeking shelter at safe houses reported that their partners threatened to hurt or did actually hurt or kill their pet.[4] Nearly a third of these women also reported that one or more of their children had also hurt or killed pets, suggesting that violence is cyclical: children who grow up in violent households, where they witness the abuse of an animal, often become abusers themselves.[5]

Public health specialist and physician Aysha Akhtar recommends that attention to cruelty to animals become a part of medical and public health education (it isn't, at this point) and that the major governmental bodies responsible for human health—including not only the U.S. Centers for Disease Control and Prevention but also the World Health Organization—take heed of the links. She recommends that veterinarians begin talking more frequently with human physicians, social workers, and public health professionals, that veterinarians receive training in identifying nonaccidental injuries, and that the profession consider mandating the reporting of abuse. Shelters need better systems for reporting abuse data and need databases that are shared with law enforcement agencies (like a uniform crime statistics database). We have strong rationale for implementing more effective intervention strategies for people *and* animals at risk, for redefining animal cruelty as part of a continuum of abuse within the family, for redefining animal abuse as a crime of violence rather than a crime against property, for recognizing animal abuse perpetrated by children as a serious warning sign of later aggressive and deviant behavior, and for viewing animal abuse as a red flag for the coexistence of child abuse, domestic abuse, or elder abuse.

If you work long enough in some facet or another of animal welfare, you will likely have heard a comment along these lines: "Why don't you stop worrying so much about animals and do something for people?" Yet research in psychology consistently tells us (and common sense confirms) that there are links between how much moral value we accord to animals and how much we accord to other humans, especially those who are different from us. Humans have a long history of "dehumanizing" certain groups of people—making them out to be less than human and more animal-like. This happened to African Americans held as slaves and to Jews, homosexuals, and gypsies during the Holocaust. Think, more generally, of the common understanding of the phrase "to treat someone like a dog." We do well to heed the warning of twentieth-century philosopher Theodor Adorno: "Auschwitz begins whenever someone looks at a slaughterhouse and thinks: they're only animals."

We should promote humane treatment of animals not only because it is good for people but also out of concern for the animals themselves. Nevertheless, it is nice to have empirical data showing links between how much moral value we accord to animals and how we treat other human beings and to know that if we raise our children to value the lives and feelings of animals, we will likely also raise children who are kind to other people and who are tolerant of human diversity and difference.

31. Heavy Petting

I want to f——k you like an animal.

Nine Inch Nails

***Warning: Some readers may find the material in this chapter offensive or disturbing. There are graphic descriptions and strong language.*

It has been around for as long as we've been etching our stories into rocks. It can be found in far corners of the world. It goes by many names: buggery, bestiality, a crime against nature, paraphilia, zoo sex, sex with animals. Yet for a practice that is so prevalent and that has such profound implications for animals, bestiality gets surprisingly little attention. To be clear: we are not talking about a few isolated incidents that catch the media's attention. Nor are we talking about the small number of violent sexual assaults on animals that get reported to the authorities and prosecuted. It is much, much bigger than that. A whole subculture of people are engaging in sexual activity with nonhuman animals (they call themselves "zoos"); there are Internet forums dedicated to sharing stories and exchanging advice; there are organized bestiality events and animal sex farms where, like a whore house, a group of animals is available for the taking. There is an entire zoo world out there, perhaps right outside your window or behind your neighbor's curtains. Although there are no precise statistics, each "zoo" likely knows, on average, about ninety other people engaged in zoophilic activity.[1] My teenage daughter reports that she knows a zoophile: he tells people that he has sex with cats and he wears a costume cat tail every day.

It is important to understand right off the bat that there is a huge spectrum of zoophilic activity, from what some people see as loving, monogamous human-animal bonds that happen to include sex, to forms of torture and zoosadism that will give you nightmares. Few people on the outside want to talk about it, and among those who do, there is considerable disagreement about whether it is a crime or a lifestyle choice, a sexual orientation or a perversion. You may be squirming a little even at the mention of this taboo subject, but no matter how hard we try to squeeze our eyes shut, it's still going to be there. And the implications for animals are huge. Bestiality or zoophilia—whatever we decide to call it—is one of the most pressing issues for all domesticated animals, including those we keep as pets. We have an entire population of vulnerable creatures there for the taking, and many people are taking.

Accounts of bestiality span cultures and time. Cave paintings from the Iron Age and the Bronze Age show men penetrating large animals with their penises and other acts of sexual congress. Religious rituals in ancient Greece often included sex with animals (e.g., men copulating with male goats; women performing sexual acts with snakes). In ancient Rome, brothels were named for the type of animals they offered for sexual purposes: *caprarii* (goats), *belluarii* (dogs), *ansenarii* (birds). Cleopatra, apparently, had a vibrator made from a box filled with bees.[2]

The eroticism of animals does not escape our notice. Children (and sometimes also adults) are fascinated by the mating behavior of animals and enjoy watching. This is often the first exposure a child has to the sexual arts, and it can be quite educational. We also—and this is perhaps even more significant—lovingly touch our animals. We pet, stroke, scratch, kiss, and whisper in their ears, behaviors that parallel intimate human interactions. For many people, the intimate interaction with an animal stops short of actual genital contact; for some, it does not.

Who are these "zoos"? From the scanty research available, the following picture emerges: the majority of zoos are male, though certainly not all; the average age is about thirty-five (many children and adolescents experiment with animals, but not all of these become serial zoos); some are married to humans, some are not; most have sexual contact with an animal several times a week. About half of active zoos use their own animal and half access someone else's. They are often consumers of pornography. They are often actively involved in animal protection.[3] Many zoos report that they are too shy to have sex with human partners, are lonely, or suffer from lack of social interaction. For some, sex with animals is preferred to and more important than sex with humans, and zoos generally seem content with their orientation.

According to Gieri Bolliger and Antoine Goetschel's typology, there are five basic sexual acts between human and animal: (1) genital acts (anal and vaginal penetration), with insertion of fingers, hands, arms, or foreign objects; (2) oral-genital (fellatio, cunnilingus); (3) masturbation (of the animal by the human); (4) frottage (I had to look this one up—to "frott" is to rub up against, for sexual pleasure); and (5) voyeurism ("mixoscopic voyeurism" or watching animals have sex, or watching another human having sex with an animal).[4] Other scholarly typologies have significant overlap, while also adding some additional types of zoophilic experience: physical and emotional attachment and affection (animal as lover), animal as surrogate for human sex partner, and fetishism.[5] To this list, I would add two more: zoosadism and zoo pornography.

Dogs are most frequently used for sex, with horses second, but a broad range of species are employed, wild and domesticated (poultry, sheep, pigs, cats, snakes, eels, you name it). Masturbating male dogs is quite common, as is licking by dogs, especially among women; anal and vaginal penetration of the animals by men, or with objects, is also common. Sex with male dogs is slightly more frequent than with female dogs.

Because I was profoundly unschooled in matters relating to zoophilic sex, I read what scholarly research I could put my hands on. There isn't much (maybe it is hard to get tenure if your research focus is bestiality). I also thought I should see for myself what the thriving Internet zoo community was up to, so I dropped in on one of the most easily accessible zoo chat rooms. I will admit, right up front, that I was uncomfortable as I picked myself a username and created an account and began reading some of the threads. (Was I being a voyeur? Was it wrong to eavesdrop on people and "research" them without their consent? Will I ever be able to erase these images from my head?) I can tell you that my foray into the zoo world only scratched the relatively tame surface of a wild and (to my mind) deeply disturbing world. I know just enough to know that what I witnessed is the mild stuff, the stuff anyone can access, that you don't have to pay for with a Visa card and that nobody is particularly trying to hide.

According to the American Society for the Prevention of Cruelty to Animals, an estimated nine hundred to a thousand zoos are actively talking to each other at any given time.[6] In November of 2013, when I visited BeastForum, it had over a million members. Among the things people were talking about in various threads were the following. In the "Howto's, FAQ's, and Technical Help" thread:

> "I've a question: How to get my fem dog to lick my dick? When I'm trying to she growls at me and don't want to lick." Answer: try peanut butter.

In the "Birds aka Fowls" thread:

> Bird for the most part are radicly different then most mamales, they share one thing in commen with a few marsupial and that is the Coclea and are very delicate far as bone structer goes. . . . If you tuely love a bird, then you'll obviously do nothing to hurt it, and you must accept the fact with small fowls, like ducks, gees, chickens, ectra it will all ways result in sever internal injury.

In a spay and neuter thread:

> i have a male fixed dog and i wanna suck his cock but how do i get it out or is it even there?

In a thread about breeders, and getting dogs from them (fear about background checks, and so on . . . what if they find out I'm a zoo?):

> I had literally no trouble getting [my Tamaskan], and to not be forced to alter them all you really have to do is say that you're going to get a tubal ligation or vasectomy instead because you found some stuff that say full spay/neuters can cause cancer and they don't ask for proof that you actually got the operation . . .

In a long thread about "Bitchsex":

> An anatomy lesson; a foreplay lesson; techniques and tips; and lots of "if she isn't receptive to you, don't force it."

In a K9 Anal thread:

> XXX.CENSORED.XXX

As you can see, at least among some of the BeastForum participants, the themes of consensuality and nonharm are present in the discussion, oddly juxtaposed with descriptions of activities that I, at least, have a hard time not viewing as exploitive.

If you are tempted to view zoophilia as a relatively benign activity, I would invite you to read around in the veterinary forensics literature. Veterinary forensics is a kind of detective work, looking at the type of injury that an animal has suffered and deducing how and when the injury was incurred. Like CSI, Animal Victims Unit. Here are a few samples from one forensics study:

> Case 3. Dalmatian. F. Sudden onset bleeding from vulva. Trauma to vagina. Dog possibly "raped" by human. Survived.
>
> Case 7. Crossbreed. F. Sharp point felt on abdominal revealed knife wound deep in vagina. Died.
>
> Case 8. Crossbreed. F. Sharp point felt on abdominal palpation was the sharp end of a knitting needle, whose other end was vaginal

in location. The 12 inch long needle had penetrated the uterine/ cervical wall. Survived.

Case 14. Crossbreed. F. Broomstick inserted into rectum up to level of liver. Died.

Case 15. Yorkshire terrier. M. Mucosa all around anus damaged and almost necrotic. Survived.[7]

Vets are certainly aware of bestiality, but as Melinda Merck and Doris Miller note in *Veterinary Forensics*, veterinary textbooks don't generally include sexual abuse as a possible diagnosis for anal, vaginal, or rectal lesions.[8] Unlike physicians, veterinarians have no reporting requirements for abuse, and most vets have minimal training in forensics. Furthermore, only a small number of zoophiles actually seek veterinary care for animals in the first place, for fear of being discovered, which is doubly unfortunate for the animals, who often suffer from infections, tears, and perforations, not to mention the emotional trauma associated with being painfully assaulted.

If the examples of violent sexual assault against animals don't yet have you convinced that something problematic is going on, then you will need to travel into even darker parts of this world (and trust me, it gets much darker). There are animal sex farms and animal brothels where people, usually men, can have their way with an animal who is kept for this purpose only. Perhaps a person is mainstream and likes dogs or sheep; perhaps he or she has a particular fetish and likes eels (tentacle porn is popular in Japan) or horses. The high-profile Enumclaw debacle in Washington State involved a sex farm where a group of men would gather for drinks and then head out to the barn to be penetrated by stallions. Some men like to penetrate poultry and often will either strangle the animal or slit her throat at the moment of climax, for the added sexual pleasure of the constrictions of the animal's sphincter as she dies. Even without the throat slitting, the sex is almost always fatal for a bird or rabbit or any other relatively small creature whose body must tear and bleed in order to accommodate the penis. Some people (we might label them "zoosadists") derive sexual pleasure from torturing or killing animals.[9] Our duck strangler would fit this description, as would the so-called cattle stabber: a person who kills cows, horses, sheep, and goats in the context of his sexual deeds. Some animals are also killed as a result of unwanted "accidents" or are killed in the aftermath of zoo sex because the perpetrator is so ashamed and disgusted.

The existence of erotic farms and animal brothels tells us that animals are being trafficked for sex, just like young women. Alan Beck and Aaron Katcher report that "in large cities one can find newspaper advertisements for 'party' dogs that are specially trained to service people."[10] There are the zoosex and farmsex and animal porn subgenres of pornography, which often take sexual exploitation to new heights by simultaneously objectifying and doing violence against animals and women. There is almost certainly pedophilic zoo porn to appeal to this demographic. And then we have, at the apex of zoosadistic awfulness, what are known as crush videos, in which small animals such as puppies, kittens, hamsters, or rabbits are crushed under a woman's stiletto heel, for your sexual viewing pleasure.

Most scholars, from my reading, consider any form of sexual contact with animals as per se abusive. Often the argument hinges on the issue of consent. Can animals consent to interspecies sex? Do they ever? It is relatively easy to know, from behavioral cues, when an animal does not want to participate (attempts to escape, cries and howls, facial expressions of pain or distress), but in the absence of overt "no" behaviors, how do we interpret the animal's willingness or lack thereof? Is silence or absence of refusal to be taken as an indication of consent? What if the animal shows signs of pleasure, interest, willingness? What if, asks Frank Ascione, "humans who practice bestiality use non-aversive operant conditioning techniques to train animals to participate in sex acts?"[11] This is not just an academic question. Much of what I saw on BeastForum has to do with what participants seemed to think of as "gaining consent." On the K9 Anal thread, this kind of language is peppered throughout the how-tos: "let the animal tell you what he or she wants"; "don't force it"; "go slowly and get them used to the idea"; "train them to like it." This blurring of the line between consent and coercion is deeply problematic.

These are difficult things to think about, if you love animals. You might say, "My owning a dog has absolutely nothing to do with those who commit acts of sexual violence against animals, just like it has nothing to do with dog fighting and puppy mills." While this might be true, I invite you to expand your awareness of animal ownership into the darker realms of animal violence and exploitation. This awareness may open the door to helping all animal victims and holding perpetrators accountable.

32. Licensed to Kill

Euthanasia is part of the well-greased machinery of the pet industry. We have, after almost a century of institutionalized killing of pets, perfected techniques for dispatching large numbers of animals efficiently, cheaply, and nearly invisibly. Every eleven seconds, a healthy dog or cat is euthanized in U.S. shelters. The pet consumer may be convinced that this euthanasia—we mustn't call it killing—is necessary and is, furthermore, an act of compassion toward individual animals and also toward the wider population of pets. But is it?

Shelter killing takes place in what Michael Lesy calls the Forbidden Zone—a place hidden within the institutions and behind the walls society has created to set death apart, to keep it from our view. "There are those among us," he writes, "who deal with death routinely. They enter the forbidden zone . . . as workers. They deal with death so we can evade it. They do what they do, armored and insulated by any number of rationalizations."[1]

I decided, with a certain amount of trepidation, to enter into this Forbidden Zone and signed up to attend a two-day euthanasia-by-injection course held at the Denver Animal Control Center during the fall of 2012. I took the class as part of my research for a book on animal euthanasia, figuring that reading about things in the abstract is different from experiencing them firsthand. I am now licensed to kill animals, and according to my instructor

Deb (not her real name), have more training in euthanasia than a graduating DVM. I find this hard to believe.

It actually took me several years to find a course being taught in my state. A euthanasia-by-injection course had been scheduled several times but canceled due to lack of enrollment. Apparently Colorado is one of a handful of states that does not make training a requirement for euthanasia technicians. If you thought that shelter euthanasia was always performed by a veterinarian, think again. A Colorado Board of Pharmacy regulation stipulates that an animal shelter worker shall not euthanize animals "unless such person has demonstrated adequate knowledge of the potential hazards and proper techniques to be used in administering such drug or combination of drugs."[2] The "demonstrate adequate knowledge" requirement is so vague that it is essentially useless. Other state laws require some training, usually either a four-hour or eight-hour course. In these several hours, a participant is expected to learn, among other things, basic canine and feline anatomy, mechanisms of action of several drugs, proper techniques for giving injections, techniques for restraint, and dosage calculations.

My euthanasia-by-injection course was taught under the auspices of the American Humane Association. Our American Humane Association–certified instructor had flown in from Kentucky. There were about twelve students in the class, representing a mix of gender and age, with no racial diversity to speak of. About half of the students worked at shelters and about half worked at animal control facilities, a pretty good reflection of where the nation's euthanasias are taking place. Some students already regularly euthanized animals but were told by their boss that they had to attend the class; others, like me, were uninitiated.

Our training guide—*Euthanasia by Injection*, published by the American Humane Association—includes all the details about how to kill animals by lethal injection as humanely as possible. As the booklet explains, although a good death isn't always possible in the shelter settings—since feral cats, fractious or fearful dogs, and basically any animal who isn't calm and quiet will be forcefully restrained—it is nevertheless the ideal toward which we should aim.

A quick look through my training manual and through the association's somewhat more detailed *Operational Guide to Euthanasia by Injection* shows how effectively euthanasia has been normalized and scientized within the shelter setting: it is all numbers and facts and veins and dosages, with nary a word about ethics. (On the bottom of the title page we read the association's motto: "The nation's voice for the protection of children & animals.") The training manual and guide were written by Doug Fakkema, one of the nation's foremost experts in shelter euthanasia and an advocate for three important improvements in shelter killing: getting all shelters to adopt lethal injection as the method of choice, making at least minimal training mandatory for anyone performing euthanasia, and convincing all shelters to use a pre-euthanasia sedative. None of these goals has yet been accomplished.

Deb opened the course with a story from her own shelter, which serves as an example of the kind of death that comes as a relief for the animal (we presume). A pit bull named Geronimo had, at the command of his owner, attacked a police officer and was being held in the shelter as evidence. The trial had dragged on for over a year, and Geronimo was becoming more and more miserable. He hated people, hated other dogs, hated the shelter. Eventually, the trial ended and Geronimo was finally put out of his misery. Euthanasia, she told us, is for the Geronimos of the world.

For the next two days, we learned about the practicalities of killing animals in shelters. We learned only "best practices" for shelter euthanasia, so we didn't learn about how to run a gas chamber or how best to shoot an animal in the field. We learned only how to kill using a lethal injection of sodium pentobarbital.

First we had to understand the various routes of administration for the injection, which involved lots of acronyms: IP (into the peritoneal cavity), IV (into a vein), IC (into the heart), and PO (into the mouth). We had a mini-anatomy lesson, looking at diagrams of splayed-open dog and cat bodies to see what target the needle needs to hit for each of these various routes of administration.

Next we had to learn about dosing. Sodium pentobarbital is a central nervous system depressant and anesthetic. The amount you inject determines whether the drug will kill or simply knock an animal out temporarily. You don't want to give too much because shelters and animal control facilities are on a tight budget and also have to account legally for all of the sodium pentobarbital used since it is a Schedule II controlled substance. But obviously you don't want to give too little and have a not-dead animal arriving at the rendering plant or landfill. To calculate the dose, you have to be able to do a little math, using the animal's weight, your intended route of administration, and other information such as whether the animal's circulatory system is compromised. Using a six-grain sodium pentobarbital solution called Fatal Plus—one of the most popular brands of euthanasia solution—for IC and IV, dosing is one milliliter per ten pounds of body weight, and for IP and PO, three milliliters per ten pounds. Within forty seconds of being injected, the animal should be "on the ground, unconscious."

Unlike the veterinary setting, where intravenous injection is favored, the route commonly used in shelters and the one we learned during the live demonstration is IC (into the heart). If you've done IC, you can quickly determine if death has occurred by taking your hand off the needle and watching. If the needle moves around in a circle, the heart is still beating; when the needle stills, the animal is dead. According to Deb, in her shelter they can bag animals once they are considered "medically dead." She once bagged a dog and was carting it through the hallway when the dog took an agonal breath; the bag moved! This can be disturbing to staff, she said. She advised that euthanasias take place in the back of the shelter, where staff and visitors are shielded from the activity. She also recommended that you laugh while you perform euthanasia on an animal because it keeps the animal happy and relaxed.

Later that afternoon, we learned about the pre-euthanasia

sedative. The sedative is generally used on fractious animals, to make them immobile so that the lethal injection can be achieved without danger to the staff. Ideally, all animals would be given a sedative first, as a kindness, but it costs a few cents more. PreMix is a mixture of ketamine and xylazine. The PreMix is injected either IM (into the muscle) or SQ (under the skin). Ketamine might sound familiar—"Special K" is a popular street drug, which is why ketamine is listed as a Schedule III substance ("high risk for abuse").

Deb recommended a twenty-gauge needle for the Fatal Plus. You clip a little patch of skin, she explained, hold the needle with the bevel up (so you can see the hole), push the needle into the vein, retract the plunger to make sure you are in a vein (you should see blood), then inject rapidly if going into a vein, slowly if you are injecting into the peritoneal cavity. If there is still a heartbeat, "something weird has happened" and you should give another dose.

Deb described her work at a shelter in Kentucky, which she said "has the worst animal welfare in the nation." During the summer months, their shelter takes in fifty to sixty animals a day. On one particular day, they took in fifty-four cats and kittens. In winter months, intake drops to around thirteen animals a day, so the save rate is much higher. They have more space, more time and energy. Intake is what drives euthanasia, she explained. The only way to reduce euthanasia is to reduce intake. It usually takes about ten years for an established spay/neuter campaign to affect a decrease in intakes. "No-kill is a myth," she told us. It is a slogan, a bumper sticker, an enabling philosophy. Her shelter does not claim to be no kill. They euthanize for medical, behavioral, and space reasons. When the shelter is full, animals must meet a much higher threshold for "adoptable." She said that a shelter in neighboring Corbon, Kentucky, had a euthanasia rate of 95 percent. When a breeder can't sell their puppies, they bring them in to the shelter and start over.

On the second day, we learned about methods of restraint, which are necessary for a great many animals in shelters and pounds. The best restraint, of course, is the least restraint. And in the ideal situation, an animal can be calmed using quiet words and stroking. They trust us, after all. Often, though, physical restraint is the only way to get a needle into an animal, whether to administer a sedative or to go ahead with the killing. Physical restraints include a stiff leash (also known as a catch pole or control pole), which is a metal wire with a plastic coating, attached to a long metal pole. You can also use a remote syringe—basically the syringe is attached to the end of a long pole. You might have, in your shelter's euthanasia room, a squeeze gate that traps an animal next to a wall and holds them still enough for an intramuscular injection of PreMix. PreMix does hurt, Deb explained, so dogs will react. There are "screamers," those that whip around and try to bite you, and the "dumb breeds" who don't really have a clue what's happening. Other restraint options include muzzles, towel collars, the Freeman cage net, and cat tongs.

Shelters are, by necessity, concerned about cost control, and euthanasia methods need to be efficient and cheap. A cost analysis matrix for euthanasia by injection compiled by Fakkema using data from a municipal animal control agency in North Carolina gives a sense of the financial impact of euthanasia. The matrix assumes that feral or fractious animals (approximately 40 percent of dogs and 50 percent of cats) are given ketamine/xylazine pre-euthanasia anesthesia; "friendly" cats are given an IP injection of sodium pentobarbital with no pre-euthanasia anesthesia; "friendly" dogs are given an IV injection of sodium pentobarbital with no pre-euthanasia anesthesia. The total equipment costs (floor safe for drugs, table, electric clippers, restraint gate) per animal are a little over one cent; labor costs per animal are about $1.38 (based on the average time of five minutes to euthanize one animal and an average hourly wage of $13.57); supply costs (sodium pentobarbital, needles, syringes, PreMix) about seventy-five cents per animal. So, the total euthanasia-by-injection cost per animal is $2.29.

On our final afternoon, we did the live demonstration, the part of the class I had been most dreading. Ideally, we should all have had an animal to euthanize, for practice. But it so happened that Denver Animal Control had only one animal on the list to be killed that day: an eight-year-old female Chihuahua with heart disease. We filed into a large supply room lined with shelves. Deb held the dog as she explained what we would be doing. The dog was shaking and wheezing and looked terrified. Even though this dog probably wouldn't have needed a sedative, Deb gave one, for the sake of demonstration. After the dog was sedated, Deb arranged the body on the gray plastic table, legs facing her, which she said she finds most convenient. We watched while one volunteer from the class performed the euthanasia. I didn't raise my hand. I couldn't. The woman who volunteered was tentative. We had chosen intracardiac injection since finding a vein would have been hard on such a small dog and the animal was already sedated. "Here?" she asked Deb, holding the needle near the heart. Deb adjusted the needle a bit and then nodded her encouragement. The volunteer pushed the needle in and depressed the syringe and then removed her hands. We watched the needle jump around in a circle and then go still. Even though I was merely a spectator, my whole body was shaking and I couldn't stop the tears from flowing. I felt responsible for taking this dog's life—a dog whose name we never even knew.

33. Rage against the Dying

I won't do what you tell me.

Rage Against the Machine

July 5, 1877, was to be a hard day. It was set aside by city authorities in New York for the destruction of unlicensed dogs in the city. Crowds gathered at the East Sixteenth Street pound as the work was scheduled to begin. It took from 7:40 in the morning until 4:30 in the afternoon to drown all 762 dogs in the East River. The next day, workers would start again with a new load of "worthless dogs" picked up by a small army of city dogcatchers. A *New York Times* story described the event:

> A large crate, seven feet long, four high and five broad, made of iron bars set three inches apart, was rolled up the aisles [of the pound], and the dogs, about 48 at a time, were dropped into it through a sliding top door. The crate was then wheeled out to the water's edge, where it was attached to a crane, elevated, swung out and dropped into the river, where it was kept submerged 10 minutes, then it was lifted up, emptied and returned for another load . . . The dogs seemed to know their fate, and most of them sullenly submitted to it; but many crouched down desperately in their corners and made a most ferocious and dangerous resistance . . . and steel "dog forks" had to be often brought into use. . . . There was one female with eight puppies that was especially hard to handle

> . . . she actually forced the other dogs within to crowd upon top of one another and give her little family plenty of room.[1]

Things have changed since 1877. Drowning has largely been replaced by other methods of killing, mostly injection. Massive round-ups of strays have been replaced by daily intake and elimination, the large crate full of dead dogs replaced by a steady trickle of bodies. Euthanasia has become assembly-line work, performed by an army of euthanasia technicians and animal control officers. The mass killing of animals is no longer a public spectacle as it was that day in 1877 along the banks of the East River. It is all but invisible to pet owners, who therefore don't have to feel discomfort or moral outrage. The slow bleed of our shelter system is one of the saddest aspects of our pet obsession.

According to Craig Brestrup's accounting in *Disposable Animals*, the first animal shelter in the United States—"where unwanted, homeless, and injured cats and dogs could be humanely housed and destroyed"—was opened in 1874 by Elizabeth Morris of Philadelphia. He says institutionalized killing might even have begun earlier, in 1858, when Morris and Annie Wahn began picking up and chloroforming stray animals.

We have certainly come a long way from the early days of institutionalized killing. We've tried, and eventually rejected as inhumane the chloroform gas chamber, the Automatic Electric Cage (an electrocution chamber), the Euthanaire (an early model decompression chamber), and the use of succinylcholine (a neuromuscular blocking agent that paralyzes the muscles while leaving pain receptors in full working order). We are gradually moving away from carbon monoxide and dioxide gas chambers, and we no longer use large metal crates for mass drownings. Yet even as the shelter system has grown and developed, and as methods for killing have become more aesthetically tolerable, the euthanasia imperative has remained constant. An amazing number of present-day animal welfare organizations buy into the killing philosophy, including the three biggies: People for the Ethical Treatment of Animals, the Humane Society of the United States, and the American Society for the Prevention of Cruelty to Animals. Killing is completely normalized and accepted as part of our cultural practice. Donors knowingly give to organizations that kill; public municipalities have contracts to kill; managers of shelters create job descriptions that include killing; people are trained, by "experts," in how to kill well.

The euthanasia issue is rich, ethically speaking, and requires attention to many complex questions. What interests do animals have in continued life (assuming we could reliably determine the answer to such a question)? In what ways does death harm an animal? Is it worse for an animal to languish in a shelter system for days, months, years or to be killed outright? These questions, and others like them, invite both empirical and philosophical answers and require a great deal of nuance, which is often lacking. It is worth noting that many of the claims made in the debate over euthanasia are cloaked as empirical statements but are unsupported by data or firm scientific understanding. For example, the claim is often made that death doesn't harm animals because they can't perceive or anticipate death and don't think about the future. This claim is meaningless (and misleading) if it is not supported by scientific data. And, in fact, it isn't. We don't know precisely how or what animals perceive as they or their friends die (and likely never will), but it very likely isn't nothing, as the growing body of research into death awareness in animals suggests . Recent research shows that a number of species of animal think about the future and make complex plans. Also, standard descriptions of euthanasia typically describe it as "rapid" and "instantaneous" loss of consciousness. This is presented as scientific fact, but we don't actually know what euthanasia feels like for animals, and never will. We can only surmise, based on our understanding of neurophysiology, and many unknowns remain.[2]

The most common justification for euthanasia of shelter animals is what we might call the better-off-dead argument.[3] Phyllis Wright, the fairy godmother of shelter euthanasia, said it like this: "Being dead is not a cruelty to animals." It is, rather, a "blessing to animals who are of no comfort to themselves or the world because they are unwanted."[4] We need to recall that she wrote these famous lines several decades ago, at a time when euthanasia of animals seemed inevitable.

Most people—even those who support shelter killing as protection against future harms—are aware that the better-off-dead line of thinking is dangerous. First of all, it is based on speculative assumptions (often cloaked as science) about what *we* think an animal might value. Second, the claim that lives are not worth living should make us immediately uncomfortable because this was a line of argument used by the Nazis to justify the killing, first, of the mentally and physically disabled and, later, of the racially impure. And finally, we all recognize the irony of sparing animals—by extermination—the continued insult of being shunned and unvalued by their so-called human companions and guardians.

In his book *The Ethics of Killing*, philosopher Jeff McMahan talks about the view that to animals suffering may matter more than death.[5] This is, as McMahan notes, strikingly at odds with the way we think about suffering and death of humans. The idea that we would kill someone simply to spare them possible suffering (even probable suffering) in the future is strange, indeed. With animals, a reverse calculus seems to be common: suffering (even the possibility of future suffering) is worse than death. This is why it is easier to get approval from an institutional review board for experiments that involve painless killing of animals than for those that involve some degree of suffering. This view is also "strikingly manifest" in the work of Temple Grandin, for example, in her various alterations to slaughterhouse design, which aim to reduce the suffering of cattle on their journey up the stairway to heaven (a.k.a. the killing floor).

McMahan says that suffering may count more, on the moral scales, for animals than it does for humans because animals may have less capacity to experience the higher dimensions of well-being (deep personal relations, aesthetic experiences, achievement through exercise of complex skills) that offset human suffering and place it in perspective. This asymmetry in animals' capacities for happiness and suffering (if, indeed, the assumptions of asymmetry are warranted) "supports the view that euthanasia is more often an option in the case of animals." But it does not support the view, "which Grandin and many others hold, that the killing of animals is generally unobjectionable if it is done without causing suffering."[6] It may be open to debate whether there are forms or degrees of suffering that are so profound that death is preferable (for human and nonhuman animals alike). But as McMahan notes, often a far more insidious line of thinking lurks behind the better-off-dead claim: that death, itself, does no harm to an animal. I'm not sure how anyone could defend this.

We could say that humane killing, although wrong, is just necessary and always will be. But this is a cop out. This assumes that we stop our moral introspection right at the shelter door. When humane movements were getting started in the mid-nineteenth century, racial segregation was mandated by law, women couldn't vote, DNA was just being discovered, and evolutionary biology was only a whisper. Most scientists did not believe that animals could feel pain, much less that animals could have complex cognitive and emotional experiences and social relationships. Maybe the killing made more sense back then, but times have changed.

I've used the term "euthanasia" (from the Greek "easy death") during these chapters on sheltering because this is the familiar vocabulary. Yet changing our language is essential, if we want to raise awareness about the costs of pet keeping and suggest to current and potential pet owners the responsibility they take on. We can discard the term "euthanasia" for what happens to animals in shelters and animal control facilities. We can call it by its true name: killing. Various people have made this argument and the reaction is nearly always the same: resistance, often anger. People seem to think that by getting rid of the euphemism, we are automatically attacking the people who do the killing—people for whom we should feel compassion and gratitude since the killing of animals causes emotional and psychological pain.

This deflection underrates the human capacity for nuanced moral reasoning and, in particular, the capacity of shelter veterinarians and euthanasia technicians to understand the tragic context within which their work takes place, and they certainly understand this better than anyone. And "kill" is simply more accurate. Although we may aspire to be humane, many of the deaths taking place in shelters are not free of pain and distress. Even in the hands of a highly skilled and compassionate euthanasia technician, animals who are resistant must be restrained, which is frightening and stressful. A euthanasia room, however sanitary, will smell like fear and death. And as those with experience on the front lines of shelter killing attest, these animals often seem to grasp exactly what is in store for them. We can condemn institutionalized violence, and work to understand the conditions that sustain it, without attacking the motives of those who carry out the killing.

Although the history of human-animal relations can be read as a tale of love and devotion, it can also be read as a long narrative of brutality and murder. Even as we have loved and housed animals, we have been at war with them and have killed them en masse when it suits our needs. But maybe war is the wrong metaphor because the animals are not armed and cannot fight back, save for the "ferocious and dangerous resistance" of an individual animal being ushered, with her little family, to her death.

Brestrup suggests that we declare ourselves "conscientious objectors to the killing, declare 'peace,' . . . let the streets fill with dogs and cats if need be, and become noisy, assertive, and unrelenting in confronting people with the moral dimensions and demands of the human-companion animal relationship."[7] Would I like to see all euthanasia technicians throw down their syringes and needles and refuse to participate? Yes. Would I like all consumers of pet culture to feel moral discomfort about their participation in an industry that results in the genocide of millions of animals a year? Yes, because only when we break the silence and truly acknowledge what is happening will we feel compelled to roar out in rage against the killing.

34. Fatal Plus

I have a purple-and-silver ballpoint pen that says Vortech on the side, a souvenir I brought home from my visit to the Fatal Plus booth at the American Animal Hospital Association convention. I suppose it was rude of me, but I couldn't help myself from asking the man at the booth, who works as a salesman for one of the most popular brands of euthanasia solution, what he thought of his job. Did he feel at all strange to be in the business of death? He seemed surprised by my question, as if it had never crossed his mind. After a few uncomfortable moments, he stammered out, "Well, if I didn't do it, somebody else would." This is how fully euthanasia is enmeshed in our pet-keeping culture: that even someone who sells euthanasia solution for a living doesn't think about the moral implications of his job.

It is difficult to find out how many pets die each year, outside of the shelter system. Nobody keeps track. There are no databases of how, where, and by what means animals are dying and no organized data collection system, as there is for humans, though such databases would provide invaluable information for veterinary epidemiologists. But we can safely make one statement about the deaths of owned pets: the vast majority of cats and dogs die by euthanasia. The vast majority of hamsters, rats, gerbils, fish, amphibians, birds, and reptiles are allowed simply to die in their cages, and euthanasia is so rarely considered for these "exotics" that few vets even know how best to perform the procedure.

There are important questions about when, and under what conditions euthanasia might be a morally appropriate choice for an ill or dying animal.[1] Euthanasia can be an agonizing moral quandary for individual pet owners, who must decide (often with little education or support) whether and when to "release" a beloved pet from suffering. As veterinary therapeutic possibilities expand, pet owners will increasingly be faced with some of the same questions that plague human patients and families who get pulled into a vortex of costly and uncertain treatment regimes.

Veterinary euthanasia is like a shape-shifter: it can be a gift, a weapon, an evasion of responsibility, or an acceptance of responsibility. It is just as ethically fraught as shelter killing, but in different ways. Within the veterinary context, euthanasia is labeled as a medical procedure and is thus ethically neutralized. As in the shelter setting, pain is separated from killing; pain is identified as the primary moral insult and killing as the remedy. Euthanasia is often included, alongside morphine and Tramadol, on lists of possible treatments for pain. Is this where it belongs?

Perhaps the most contentious moral questions surround the issue of convenience euthanasia, where a veterinarian is asked to kill by a pet's owner for reasons that have little or nothing to do with the pet's well-being. There are no data on how common convenience euthanasia is, but I've never talked to a vet (and I've talked to a lot of them) who hasn't faced the issue on multiple occasions. And nearly all of them find these requests distressing, even repulsive. Vets are healers but are being asked to kill. I think we would all like to know: How have we ended up in this place, where a person can request the death of a healthy animal, simply because they feel inconvenienced in some way by the creature?

Here in the United States, veterinary euthanasia seems to be something of a sacred cow. Even the suggestion that we cut way back on euthanasia practices strikes people as revolutionary (in a bad way). Could we envision a pet-keeping culture in which euthanasia didn't exist, or was extraordinarily rare? Actually, yes. All we need to do is look at other countries around the world. There are pet-keeping cultures who don't have surplus animals and don't kill the unwanted or kill in much, much smaller numbers. There are also countries where veterinary euthanasia at the end of life is rare, is considered morally objectionable, and where many pets die a natural death. Broad-scale euthanasia is not the only possibility. Nor is it the best.

35. Eunuchs and Virgins

Spay/neuter, like euthanasia, is an unquestioned part of our current cultural narrative of pet keeping in the United States. Pet owners are told by authorities that it is absolutely necessary to have their animals "fixed." This is why I did it. I've felt sad for each of my animals, like I've deprived them of something special, perhaps even sacred. But I've also told myself: there is no other way. (Fixing an animal involves the surgical removal of the gonads—the organs that produce sex hormones.)

The pressure to sterilize is intense. Spay/neuter is the mantra of veterinarians, the nation's humane organizations, and the majority of animal activists. The American Veterinary Medical Association, for example, says that sterilizing is essential to responsible pet ownership. If the goal is really to have no reproductively active animals in the hands of private pet owners, then we are making good progress. According to the Humane Society of the United States, 83 percent of owned dogs and 91 percent of owned cats are spayed or neutered.[1]

Three reasons are offered in support of the campaign to spay/neuter America's dogs and cats: (1) our animals will be healthier, (2) our animals will be better behaved, and (3) we must spay and neuter as many animals as possible to control pet overpopulation and avoid unnecessary killing. At first look, these seem simple enough. Yet none of these is without controversy, and even some veterinarians and animal activists are starting to wonder if the whole spay/neuter campaign has been oversold.

The balance of benefits and detriments to individual animals is hard to gauge, and there seem to be genuine trade-offs. Sterilization decreases the risk of death from some causes (e.g., infectious disease), while increasing risk of death from others (e.g., cancer).[2] In female cats, incidence of mammary gland neoplasms is reduced, while incidence of obesity is increased. Intact male dogs are more likely to develop testicular neoplasms but less likely to suffer from obesity, cranial cruciate ligament rupture, and prostatic neoplasms.[3] Neutered and spayed dogs have higher risk of some cancers (lymphoma, prostatic adenocarcinoma, transitional cell carcinoma) and lower risk of others (mammary gland tumors, testicular tumors).[4] Research suggests that sterilized dogs live longer than intact dogs (sterilized humans live longer than intact humans, too), but scientists don't yet have a good picture of why.

Private veterinarians typically recommend waiting until about six months of age or until after the first estrus cycle for female animals. But shelters don't want animals to leave the door without being desexed, so they sterilize early—a decision based less on science than on practicality. Is pediatric spay or neutering risky for dogs or cats? Again, the answers are not obvious. Anesthesia is somewhat riskier for younger animals, but current drugs and methods are pretty safe, so this isn't really a significant added risk. Some veterinarians believe that too-early spay and neuter deprives the animal's developing body of the hormones that shape sexual development and that reducing or eliminating these hormones may cause permanent harm. For example, early spay/neuter may delay the closing of growth plates, leading to increased bone growth. Abnormally long bones might alter joint angles and increase the possibility for arthritis, hip dysplasia, and cranial cruciate ligament injuries. On the other hand, "having an intact, but unused, reproductive system may create as many, albeit different, problems as the lack of one," writes veterinarian Myrna Milani.[5]

We've been told that neutering and spaying creates an animal with fewer behavioral problems. Let's be clear: these purported behavioral benefits may make our animals *better pets*. We want submissive, obedient animals; we don't want dogs and cats trying to get out and roam the neighborhood looking for sex partners when they should be at home sitting by our feet; we don't want to have to deal with the mess of female estrus or male scent marking—entirely natural behaviors, we might add, but ones that make pet keeping harder work.

Does desexing actually accomplish the behavioral modifications that we desire? As with health benefits, the research is not conclusive. The clearest positive advantage seems to be in relation to tom cats. The behavior of intact male cats, particularly urine marking in the house, is bothersome to humans and is generally successfully "treated" by sterilization. If these are cats who would have been relinquished to a shelter for peeing in the house—and this is a likely scenario—then there seems a good argument for sterilization. There is evidence that surgical neutering *may* reduce some hormone-dependent and undesirable behaviors in male dogs such as roaming (though a tall fence can also be effective). Whether castration leads to decreased aggression in male dogs is unclear. Although there seems to be a correlation between neutering and aggression among shelter dogs, research suggests that the correlation is less clear among owned dogs. Furthermore, while neutering may decrease aggressive behavior between male dogs competing for sexual "territory," there are many forms of and motivators for aggressive behavior, and we have too few data at this point to support blanket statements like "neutering reduces aggression." Behavioral benefits of spaying female dogs and cats are even less clear. Milani argues that connections between intact status and certain behavioral tendencies are tenuous and the putative benefits of sterilization are likely overstated.

Are there behavioral downsides to sterilization? Researchers don't know as much as they would like about the non-reproductive effects of hormones on an animal's physiology and behavior. An increase in reactivity has been reported in female dogs after ovariohysterectomy, a procedure in which both the uterus and the ovaries are removed. There is some evidence that spayed female dogs may develop urinary incontinence and that castration of male dogs is linked to hyperactivity and canine cognitive dysfunction.[6]

Spay/neuter has been the orthodoxy for about four decades, and it is fair to say that the spay/neuter ideology has been successful: people are largely convinced, more and more pet owners are doing the "right" thing, and spay/neuter clinics are proliferating. One of my veterinary friends performs so many neuter surgeries on cats that she makes little gonad necklaces for her officemates. During these forty years, the number of shelter killings has dropped from a staggering twenty-three million to a merely horrific three to four million.[7] Although the demographics of dog and cat populations are complex, and correlations between spay/neuter and numbers of shelter animals killed are not straightforward, there is no question that spay/neuter programs are reducing the number of unwanted pets and shelter killings.

Others argue, however, that pet overpopulation is not inevitable, nor is control of numbers of animals necessarily linked with aggressive spay/neuter policies. Responsible pet ownership can, according to this line of argument, be as effective as spay/neuter and far less costly for the animals. As evidence, they point to European countries who do few spay/neuter surgeries, yet maintain low numbers of unwanted animals. How? Few animals are allowed to run loose, and so there are few unplanned pregnancies. In stark contrast to moral attitudes toward spay/neuter in the United States, removing the sex organs of pets is strongly discouraged, except in cases of medical necessity. In Sweden, only about 7 percent of pets are desexed, and in Norway, it is currently illegal to desex a healthy animal.[8]

Furthermore, "pet overpopulation" may be a faulty description of the state of affairs. The problem may not be that we have surplus animals so much as that people who want to procure an animal don't routinely adopt from a shelter. According to the Humane Society of the United States data, only about 20 percent of dog owners and 26 percent of cat owners report having acquired their animal from a shelter, despite the fairly high-profile coverage of the puppy and kitten mill problem (see chap. 37, "The Shelter Industry").

Putting questions of surplus and overpopulation aside, there is one final and rarely discussed (and disturbing) reason to spay or neuter animals: it might give them a small measure of protection from human sexual predators, who generally look for unaltered animals. Finding an intact animal is less important if the target is a female, since the act of penetration (anal or vaginal) will be the same, and it doesn't matter if the animal is sexually aroused. But if one is interested in having a male animal who will respond to frotting or who will mount a human, then I gather (from eavesdropping on Internet chatroom discussion among zoos) that intact is the way to go (see chap. 31, "Heavy Petting").

Veterinary ethicist Bernard Rollin argues that because animals enjoy sex as much as we do, we ought to look for ways of controlling reproduction that don't involve complete removal of sex organs and sex hormones.[9] One possibility is to choose the least invasive and least aggressive surgery: vasectomy for males rather than castration, and either tubal ligation or ovariectomy, or removal of the ovaries, for females, rather than removal the ovaries and uterus (ovariohysterectomy). Ovariohysterectomy—the more invasive procedure—is standard in the United States. In Europe, ovariectomy is the preferred method of sterilization, and there is no reason the United States cannot follow suit.[10]

The Alliance for Contraception in Dogs and Cats, while advocating for widespread control of dog and cat reproduction, is promoting alternatives to surgery. Unfortunately, good oral contraceptives for female dogs and cats are not readily available. There are several products used to postpone an estrus cycle, used primarily by breeders or in show dogs or hunting dogs, but the drugs are intended for short-term use and have serious enough side effects that they are not a good option for general use.

As far as alternatives to castration go, the picture looks a little brighter. One promising new technique for canine neutering is called Zeuterin. As I write, Zeuterin is just becoming available in the United States. Zeuterin was developed by Ark Sciences and involves the injection of gluconate and arginine (what you find in the cold medicine Zicam) directly into a dog's testicles. On injection, the compound diffuses and kills spermatozoa. Although Zeuterin renders the dog sterile, it only reduces testosterone by 50 percent, whereas surgical castration eliminates testosterone entirely. If testosterone plays a protective role in canine health, this is a benefit. Furthermore, chemical castration is less invasive and less painful than surgical castration and involves use of a sedative, but not anesthesia, further reducing risk to the animal. Chemical castration is also cheaper than surgery.[11] Because chemical castration leaves the family jewels externally intact, it will appeal to that segment of the pet-owning demographic who wants a dog to feel and look like a real man. According to a webinar I watched (sponsored by Ark Sciences), reluctance to remove a dog's testicles is particularly prevalent among people who use dogs for work (herding or hunting or protection) and who worry that castration would make the dogs less effective at their jobs. Whether this is a real risk is up for debate, but it is a common perception. Another possible alternative to surgical castration for dogs and possibly also cats (but not yet approved in the United States) is an implant of a drug called Suprelorin, which neutralizes gonadotropin-releasing hormones, resulting in inhibition of testosterone production.

People are trying to bring nuance to the issue of spay/neuter by separating some of the pieces that are generally lumped together. For example, when we talk about spay/neuter campaigns, we generally lump together cats and dogs. Yet population dynamics are very different for cats and dogs in the United States. For example, there is almost no interstate transfer of cats, whereas dogs are actively being transported across state and international lines for rehoming. We have large populations of feral cats but relatively few feral dogs. So our approaches to desexing and population control need to differ according to species. Different parts of the country also have different pet population dynamics, and there may be some geographic areas in which aggressive spay/neuter should be a high priority and other areas in which it may be unnecessary or at least more discretionary.

Dynamics are also different in shelters than they are among private pet owners. Shelter animals are being treated as a group, not as individuals. Blanket sterilization is rather like vaccination: it is not primarily aimed at the welfare of the individual and in fact may involve some small risk or cost, but it promises significant benefit for the wider community of animals and people. Shelter vets are more akin to public health officials than private physicians. Within the private setting, the vet's primary moral obligation is to the individual patient, and the calculation of risks and benefits of sterilization might look quite different. If there are benefits to male or female dogs or cats in retaining their reproductive organs, it becomes ethically problematic for veterinarians and humane organizations to essentially force—through shaming, coercive presentation of information, peer pressure, and bullying—individual pet owners to compromise their individual animal's well-being for a hypothetical risk that the dog or cat might accidentally reproduce.

There is a tendency to oversimplify the issue of spay/neuter and to promote the essential benefits without recognizing that our animals do suffer some harm, even if it is only the harm of deprivation—the harm of having their sexual and reproductive experiences stolen from them. It is possible to take this argument to the extreme and assert that we should never interfere with something as basic as sexuality and reproduction. Good stewards would allow their animals to exist in a "natural" state. The problem here is that our companion animals have no "natural" state; as domesticates, they are artifacts of human manipulation, and human control over the processes of reproduction is at the heart of domestication. As Karla Armbruster notes in her essay "Into the Wild," we cannot simply hand control for reproduction back to our companion animals; this would be an abrogation of our responsibility to them. But we owe it to them to acknowledge their losses.[12]

We could take an even darker perspective on spay/neuter campaigns. Perhaps one of the reasons that spay/neuter is necessary right now is because of larger forces at work in the pet industry. Commercial breeders, large-scale and small, want to keep breeding in their own hands, to control the flow of product, as well as its price and quality. They want us—the nation's pet owners—to desex our animals so that when we need a replacement, we must buy another from a "reputable" manufacturer. This keeps breeding under the control of those who profit from it (see chap. 36, "Breeding Bad"). It also reinforces the status of animals as commodities (see chap. 39, "A Living Industry").

These more radical challenges to spay/neuter are offered as experimental lines of thought. Broad-scale sterilization of dogs and cats may be a reasonable, if costly, solution to our current set of problems. But the solutions and the problems both need further exploration.

36. Breeding Bad

In intensive pig-farming operations, gestating sows spend their entire pregnant life in a small square of space, surrounded by metal bars and with slatted concrete underneath so excrement can fall onto the floor below. Gestation crates are considered among the cruelest of modern farming practices. If you've never seen a picture of one, go troll around on the Internet and find some images. Now, do a search for "commercial puppy breeders" and see how the production of puppies compares with the production of piglets.

You'll notice some striking similarities: the bodies crowded together, yet unable to interact; the look of despair in the eyes of the animals. You'll notice some differences, too. For one, the gestation crates look relatively clean, while the dogs and their cages are covered in filth. If you balk at the idea that dogs could be bred like pigs, I'm sorry to disabuse you. Dogs and pigs alike are treated as breeding livestock. The animals are made to have as many young as possible. The babies are taken away at a young age so that they can be sold and a new breeding cycle can begin. The animals never have real sex—that is, sex when and with whom they choose; rather, females are tied up so they can be mounted or, more often, they are artificially inseminated. In commercial breeding operations, and also in many small-scale or backyard breeding outfits, dogs are treated, like the sows and their piglets, as units of production and their sole function is to bear young for profit. All they do is bear one litter after another, until (usually at the age of four or five) they are spent. At which point they are no longer of value and are killed. Taken together with the spay/neuter picture, what we have is rather bizarre: an enormous population of eunuchs and virgins, and a small population of dogs who live their entire meagre existence as breeders, as part of a puppy production line.

Information about commercial breeders—the worst of them are called "puppy mills"—is readily available. It is worth a quick review, though, particularly if you are one of the many people who believe the puppy mill problem is a thing of the past. Mills are doing just fine, particularly in the so-called puppy mill states: Missouri, Nebraska, Kansas, Iowa, Arkansas, Oklahoma, and Pennsylvania. Estimates are that 90 percent of puppies sold in pet stores each year come from commercial breeders, as do a great many of the dogs sold over the Internet, at auctions, and through newspaper advertisements. The American Society for the Prevention of Cruelty to Animals estimates that there are anywhere between two thousand and perhaps ten thousand commercial breeders at work in the United States.[1] Some of these are operating legally, as USDA-licensed facilities. Others are small enough to escape regulation or simply function illegally. Some operations house only a few breeding females and puppies, whereas others may house thousands of dogs. Whether a breeder is operating legally doesn't matter all that much, in terms of animal welfare. The miniscule number of USDA inspectors charged with overseeing commercial breeders and Class B animal dealers guarantees even legal operations a large measure of freedom and privacy.

Reputable breeders will often sell straight to the customer—and, in fact, some insist on meeting the prospective parents of their puppies. But many breeders rely on a network of animal brokers to purchase the puppies and sell them to pet stores or online. Puppies are taken from their mother at a young age. Eight weeks is supposed to be the minimum age at which puppies can be separated from their mother, but they are often weaned and sold much younger, so that they can be delivered to pet stores at the optimal age of eight to ten weeks. They are transported by truck or van to a broker, who then arranges sale and transport to pet stores or auctions. Before they even arrive at their final destination, many of the puppies will have died of illness or malnutrition.

One of the largest commercial dog-brokering operations in the United States is the Hunte Corporation, located in Goodman, Missouri.[2] The Huntes got their start in the puppy production business selling Sundowner Kennels, specially designed units that house rows upon rows of animals for breeding. (If you want to start a puppy mill of your own, you can find lots of used Sundowners on Craigslist.) Hunte's facilities cover eighty thousand square feet, with plans to expand to a hundred thirty-five thousand square feet. They proudly claim to sell over a hundred thousand puppies a year, both in the United States and overseas. Not surprisingly, Hunte has been the focus of attention for several animal advocacy groups. Undercover footage inside Hunte's daily operations, collected during an intensive investigation by the Companion Animal Protection Society, is well worth watching.[3]

I'm focusing here on dogs because this is where almost all of the research and exposés lead us. But of course puppies aren't the only pet animal being bred and brokered and sold for profit; they are just the most high profile. There are kitten mills, too. And rabbit mills. And the many other animals who we keep as pets—the rats, hamsters, and geckos—don't just materialize out of thin air; they come from a mother somewhere, who has been intentionally bred so that humans can make a profit selling her babies (see chaps. 38, "Cradle to Grave," and 39, "A Living Industry").

There is wide diversity among animal breeders, with varying motivations and varying standards of animal welfare, and it would be wrong to place them all into the same category. There are family pet owners who have no intention to breed, per se, but want to provide their animal with the experience of giving birth and raising young at least once before they are sterilized; there are family pet owners who have a litter of puppies or kitties so that the children can witness the "miracle of birth" (but they are shielded from the "miracle of death" that breeding leads to); there are breeders who simply put the family pet to work to help support the family and pay for dog food; there are hobby breeders who just love a particular breed of animal and enjoy promoting it; there are money-oriented breeders, small scale and large, who don't care much about the animals themselves and perhaps even feel distain; and of course there are many accidental breeders.

There are certainly responsible breeders, large scale and small, who care about their animals, who treat them with love and respect, and who aim to produce healthy and well-socialized offspring and ensure that they are placed into good homes. Nevertheless, some animal advocates would say that "responsible breeding" is an oxymoron and that even the most reputable and humane breeders should be put out business. Huge numbers of surplus animals are languishing in shelters, and until there are no more homeless dogs and cats, people should all be adopting from the shelter. Every puppy or kitty bred is one less shelter dog or cat who will find a home and who may be put to death. This is pretty compelling, on its own. But the ethical problems with breeding run deeper.

Another ethical concern with breeding—and one that has received very little attention, perhaps because it smacks of sentimentality—is the fact that we take babies away from their mothers. Sentimentality aside, there is solid evidence for acknowledging that the mother-infant bond is a basic biological drive: mammalian species have evolved strong parental caring behaviors. Separation is emotionally painful and stressful for baby and mother alike. We have no reason to suppose that this mother-baby bond is any less strong in dogs or cats or rats or other mammals than in humans, yet taking babies from their mothers is a regular aspect of pet-keeping practices.

This past Christmas, puppies seem to have been a very popular gift in my small town. For the couple of weeks after the holidays, I regularly ran into people out walking with an adorable new puppy which, of course, I would have to stop and meet. (Puppy breath!) Even as I delighted in the puppies, I couldn't help but feel sad for the mother dogs, who lost their litters, and for the puppies themselves, who generally look dazed by their new circumstances.

It is by now well-known that selective breeding and the maintenance of breed standards—what Michael Brandow calls "the cult of pedigree"—has been damaging to dogs.[4] (There are far fewer concerns with cats, mainly because a much smaller percentage of cats are purebred—only about 15 percent.) The problems with pedigree were first brought to public attention in 1990, with Mark Derr's *Atlantic Monthly* exposé "The Politics of Dogs." As Derr reported, breed standards have long focused on appearance, rather than on behavior or on overall health, and the result is a nation full of dogs who must live with disability and discomfort. Think of the brachycephalic dogs like bulldogs, Pekinese, pugs, and boxers, who have been bred with flattened noses and bulging eyes. The shape of the skull has been altered to such a degree that their brains have, by necessity, shrunk. Veterinary offices are filled with dogs who can barely breathe, who can never run to their heart's content because their joints are malformed, who have eyes that are constantly infected. Dogs are developing crippling hip dysplasia or arthritis before they even reach middle age. Others are born deaf, suffer infections in their oversized skinfolds, live with inherited blood disorders, or are killed just after birth because they are "defective." Closed stud books make genetic defects almost inevitable. (A stud book is a registry of animals descended from foundation stock. To be a purebred, both parents of offspring must belong to the relevant breed registry. Closed stud books allow no new blood into the line, which helps to maintain a narrow set of breed characteristics.)

Many kennel clubs and dog shows continue to require conformation to standards that compromise the well-being of animals, such as the sagging hips of the German shepherd, despite overwhelming evidence that lowered hips are a recipe for disaster. To support continued breeding, breeders and kennel clubs often invoke tradition, saying that it is tragic to "lose" a particular breed like pug or boxer or shepherd. But even forty years ago, these dogs looked radically different, so "tradition" must be understood as quite arbitrary and rapidly changing. As with so many aspects of pet-keeping culture, economic profit seems to pull against animal welfare. Kennel clubs make money from registering purebred dogs, and the financial incentives push kennel clubs to ignore problem breeders, puppy mills, and genetic defects. Many dogs who were originally bred for functional work—herders, pointers, terriers—have lost their capacity to perform these tasks. Their "work" now is to look pretty. As Derr remarks, dogs bred for show become little more than a piece of sporting equipment used to compete in a game. Derr's essay also suggests just how quickly the tide can turn against a certain breed of dog. For example, the shar-pei went from being, during the1980s, one of the nation's most sought-after and highly prized (and priced) dog breeds, to becoming a "genetic disaster"—with skin infections and familial shar-pei fever, among other things—and filling up the nation's veterinary offices and finally the nation's shelters. As more people become aware of the problems of inherited disorders and inbreeding, some breeders and even a few kennel clubs are bucking tradition and are allowing the health of the animals to guide decisions about which genes should be retained. The United Kennel Club, to give one illustration, requires all registered dogs to have health certification. This is a step in a better direction.

Ultimately, though, the cultural obsession with "purebred" dogs could usefully be swept into the dustbin of history. The assumption that purebred dogs are "better" needs to be answered with the question "better for what?" If what people are seeking is status or a blue ribbon or the fulfillment of some particular homespun notion (the perfect family, complete with golden retriever), there are plenty of ways to achieve these things without harming animals in the process. Being a "half-breed" is no longer an insult for a human; it shouldn't be an insult for a dog either. In fact, having a diverse racial background makes a person really interesting, and so, too, with dogs. Let's learn to love the mongrel and the mixed breed and to celebrate the canine melting pot. This is the American dream, after all.

37. The Shelter Industry

Many people are complacent about the fate of animals in our society because a great big system of lovely shelters exists to take care of unwanted animals and to find them new forever homes. So it isn't really a big deal if you are moving and can't take your animal to the new apartment, or if your child has outgrown his interest in the puppy and the family is just too busy to give her enough attention. She will get a new home before long. And in the meantime, a team of nice, animal-loving people are on hand to care for your friend. And so the fantasy goes.

This fantasy fuels the single biggest moral problem with our pet-keeping practices: millions of abandoned, unwanted, abused, relinquished animals languishing in our nation's shelters, pounds, and humane societies. This is the cost, in real lives and real suffering, of our culture's pet obsession. Even if you, as an individual pet owner, would never conceive of abandoning your companion, you are not absolved. None of us is innocent. Society as a whole is responsible for "the shelter problem," which is actually not a single problem but a complex array of interrelated problems arising from the way we approach pet keeping: animals are viewed as products to consume at whim. Too many pet owners feel only weak and transitory responsibility for the animals they take under their care, and when they are finished with their animal, or when the animal loses its shine or bites back after one too many beatings, they have few qualms about disposing of the creature.

Somewhere between six and eight million dogs and cats and other sentient creatures pass through the shelter system each year. Some lucky ones do, indeed, leave for a new and forever home; some cycle in and out and in again (typically those who were deprived of the socialization and teaching that is required of well-behaved pets or who have "bad temperament"); and many—at least half—leave the hard way. Taking a middle of the road estimate, about three million companion animals are killed every year in our nation's shelters. That's 342 every hour. In the time it has taken you to read this paragraph, a life will have been extinguished.[1]

To call it a "shelter system" suggests too much consistency and standardization. A diverse patchwork of people and organizations try to manage the flow of surplus pets. The Humane Society of the United States is not, as many people assume, a huge umbrella organization with satellite humane societies in cities all over the country. It is an advocacy group, not a shelter. It has no direct relationship to my local Longmont Humane Society, much less to the Denver Dumb Friends League or the local greyhound rescue. Many shelters are small 501(c)(3) nonprofit organizations, which run under contract with a city and have to do a lot of fundraising to survive. Other shelters are run by municipal animal control, under the purview of a city or county government—these are often still referred to as "the pound." Besides these brick-and-mortar shelters, an increasing number of loosely organized foster home networks and breed rescues and small-scale sanctuaries have sprung up over the past decade or two. This sheltering system is constantly evolving, in response to changing demographics and the dynamics of pet supply and demand as well as other factors such as the use of the Internet and the increasingly common practice of interstate and international transport of dogs.

Among those places that take in lost, unwanted, and stray animals, some are horrible beyond belief. Some are quite nice and offer better living conditions than many homes. Most probably fall somewhere in between. If you have never been inside a shelter, take an hour out of your day and walk through one. Make eye contact with those dogs who are willing, watch their eager tail wagging, or see the one who fails to even lift his head for you, whose time there has drained him of happiness and hope. Watch the cats sleep away their days with nothing to do. And don't forget about the small animals—the rabbits and rats and gerbils and sundry other warm- or cold-blooded creatures the shelter is willing to accept. And don't visit just one shelter because there is huge variability. You could visit the Boulder Humane Society and come away feeling okay about things. It is a nice place, well-funded, overflowing with almost as many staff and volunteers as animals, and it sustains high adoption and low euthanasia rates. I've been in other shelters in my area that made me unbearably sad. One is a huge warehouse of chicken-wire kennels, with a constant cacophony of barking and a pungent smell of feces and urine. There are far more animals than people, and the animals look stressed out and depressed. I'm sure, based on my geographical location, that I have seen nothing of the horrors that some shelters in some parts of the country have to offer.

Increasingly, homeless dogs (and to a lesser extent cats) are being cared for by a loose network of individuals running small-scale rescues or sanctuaries or foster services. Often these rescues serve as a place of last resort—they take, from a local humane society or shelter, the animals whose time is up, who have been placed on the dreaded PTS (Put to Sleep) list. There are perhaps hundreds of thousands of dedicated people who are saving these last-chance animals, and each one of them deserves a medal.

Yet it is important that the public understand that there are no legal or standard definitions for "shelter," "rescue," or "sanctuary." Just because a place is called a "rescue" or a "sanctuary" does not guarantee that animals are being lovingly cared for in a warm, clean home. Every couple of weeks I come across a news report about a so-called rescue or sanctuary that has been raided by animal control or whose owner has been charged with animal cruelty because of the abhorrent conditions in which the animals are kept. For example, in February of 2014, Puppy Patch Rescue Group in Morristown, Tennessee, was raided and one of Puppy Patch's owners was arrested on sixty-three counts of animal cruelty. Dogs, cats, and fish were taken from the residence. The animals were being kept in small and filthy enclosures, often without food or water.[2] The Animal Rights Foundation, a rescue group in Beachwood, Ohio, made the news when the owner committed suicide. She was found in her car in her garage, car engine still running. She brought thirty-one dogs, mostly puppies, along with her. All the dogs died but one, a small pup who squeezed out of the car and found a small hole in the garage for air.[3] (I'm not sure whether to read this as a story of animal cruelty or as a cautionary tale about the despair that trying to rescue unwanted pets can engender in empathic people.)

One of the saddest things about shelters is the story behind each individual animal who finds herself suddenly unmoored from her familiar life. Many animals in shelters have been picked up off the streets by animal control—having been lost or set loose or having escaped from a house or yard. Only about 16 percent of lost dogs and 2 percent of lost cats are claimed from shelters by their owners.[4] About a third of animals entering the shelter system are there because their owners brought them in. Some shelters have surrender forms and ask for a small fee. They try to elicit from the owner why the animal is being relinquished in an effort to better understand how to stanch the flow of bodies. Compelling and heartbreaking stories can be found among these surrender documents. Sometimes a person falls ill and cannot care for their animal, or an elderly person is moved to a nursing home or dies and the family doesn't want the animal. Yet if you read through the stated reasons for surrender, you will likely find yourself saddened at the general level of irresponsibility and lack of commitment to our animals: "moving, can't take with," "boyfriend doesn't like," "allergic," "barks too much," "sheds."

Diane Leigh and Merilee Geyer have spent years volunteering or working in shelters and beautifully capture the essence of the tragedy in their book *One at a Time: A Week in an American Animal Shelter*. They write,

> There is a moment, when the paperwork has been completed, and the animal is being handed over to shelter staff . . . if you watch carefully, you can sometimes see the exact moment when the animal comprehends what is happening, when he finally realizes that his guardian is leaving and he is staying; the exact moment when the confusion in his eyes is replaced by understanding, and then turns to panic, desperation. Sadness, that will turn to grief as the days unwind, while he waits for another chance that may or may not come.[5]

Most surrendered animals haven't been in their home for even a year. Half are not spayed or neutered. Some have never seen a vet. Most surrendered dogs have no training, and some have only lived outside. Behavioral problems are common. One family I know offers a good example of what happens. They are "serial adopters" and could be a poster family for the shelter system. They have, in the ten years that I've known them, adopted at least five dogs—I can't even keep track. The mom has a big heart and loves animals, and when she doesn't have one, she yearns for one. But once they adopt a dog, they realize, time and time again, that they are just not equipped to handle the responsibility. The last dog they adopted was a three-month-old lab/husky mix. He was a sweet, rambunctious dog, and full of life. The dog lived in the backyard, never went on a single walk, never got house trained, and was taught none of the skills that he needed to make him a nice pet. The husband sort of likes dogs, but he has a serious dog-poop phobia, which means that the full-time working wife has to do all the cleaning up, which causes marital tension. After about six months, the dog is too big and is knocking over their youngest daughter (who has been taught nothing about how to interact appropriately with an animal). And the dog poops way too much! After a few tears are shed, the dog goes back to the humane society. He is no longer a puppy, he is unsocialized and untrained, and it will be considerably more difficult for him to get placed in a new home. About six months later, the serial adopters go back for a new dog.

The research on relinquishment suggests, note Leigh and Geyer, "a basic underlying dynamic: that the decision to acquire an animal was made casually, without much forethought and planning."[6] And so one of the things shelter managers and animal advocates have been trying do is to help people make better decisions about adopting and especially, perhaps, *not* adopting a new pet. Shelters have also been working hard to increase retention of adopted animals, particularly by focusing on education and better matchmaking. More and more shelters are offering or even requiring "pet parenting" classes for new and prospective adopters, where people learn what to expect from an animal, what their animal needs to learn in order to be a pleasant and well-socialized member of society, and how to have more realistic expectations about what their responsibilities to this creature will be.

Shelters are also offering training and support once an animal has gone home with someone. Shelter behaviorists make themselves available to help troubleshoot behavioral problems and offer advice and support. One recent study found that dogs are 90 percent less likely to be surrendered when their guardians have access to behavioral advice.[7] Shelters are also exploring how to make sure potential adopters understand what they are committing to financially, for example by offering explicit "calculated lifetime expenses" of owning a dog (ballpark of $20K) or a cat ($10–15K). People interested in adopting a guinea pig for their child will learn that it will likely cost them about $635 a year, far more than most parents anticipate.[8]

My local humane society has very high adoption rates, but despite the high turnover, some animals—those who are old or ugly or who have behavioral challenges—are passed by and can wind up living in the shelter for long periods. There are dogs and cats who have been there for several months or sometimes as long as a year or two. No matter how nice a shelter is, life there is scary and stressful for animals. They are exposed to high levels of noise and activity, confined to a small space from which they cannot escape, and are often alone—despite the constant flow of human traffic and the presence of many other animals. Dogs and cats (and certainly other animals, too) can develop "kennel stress," which can be devastating and life ending.

One obvious solution is to design shelters in ways that mitigate stress and that cater to the needs of the animals. For example, dogs are increasingly kenneled together in pairs or small groupings and are provided with communal play areas and stimulating activities. At my local humane society, every dog gets at least two outside walks a day with a volunteer dog walker. Cats have somewhat different environmental needs. Most obviously, they benefit from being acoustically protected from barking dogs and other loud noises. They like places to hide and climb and sleep in the sun and opportunities to interact with or haughtily ignore other cats. Creating humane shelters, of course, requires ample community resources and support and, thus, a strong commitment to animal welfare.

Attitudes toward sheltering are evolving, as reflected in a changing vocabulary. It used to be called the pound (literally, "shut up" or "enclosed" from Old English); then it became the shelter; it is now a campus. Campus sounds nice, like doggie and kitty college. Consistent with the campus theme, shelters increasingly emphasize education—both of the animals and of adopters and the broader community. Rescue is increasingly referred to as "adoption," which is apt because it emphasizes long-term stability and draws the obvious analogy between pets and children. Animals are not abandoned by their owners—they are "relinquished" or "released." Animal control is increasingly becoming "animal services." Animals are not destroyed or killed; they are euthanized. This retooling of the language is great, since it reflects an even stronger commitment to improving the welfare of the animals stuck in the system. It is increasingly euphemistic, too, and that's the downside. It makes it easier for us to feel good about what's happening and to think that all is well.

There is a built-in dilemma with shelters. You can be like the Boulder Humane Society, which is clean, bright, full of caring people and offer about as good a life as can be had for shelter-bound animals. The kill rates are low. It is still a sad place to visit or volunteer—so many sweet and lonely creatures in need of a loving home, and some still being killed. But you can feel wistfully sad rather than horrified-sad. The problem is, the public is let off the hook to a certain degree. We can feel warm and fuzzy inside because the community really cares about animals. We can think to ourselves, "It's not so bad to be a shelter animal."

Shelters often rely on public donations, so they have to constantly do outreach in the community. They may feel compelled, by their need to remain solvent and sustainable, to shield the public from the worst and to garner support through some mild "humane washing." They keep a positive spin on things, choosing images for websites and advertising material that will best bring in funding. They highlight the stories of hope and resurrection, not the stories of all those animals who have been killed this month. Shelters certainly don't have euthanasia technicians doing their work out in the lobby or in a glass-walled room. In the meantime, shelters are codependent, reinforcing the public's capacity to look the other way and think things are just peachy.

Are shelters part of problem they seek to address? Does the shelter industry enable our dysfunctional relationship with animals? In some ways, yes. The fact that it is referred to, by insiders, as "the shelter industry" might send up a red flag because an industry is something that seeks to grow or at least remain sustainable over the long term. Ultimately the goal, it seems to me, should be empty shelters. But a shelter without animals, or with only three or four animals, cannot thrive. Shelters rely on having a high census and adopting out as many animals as possible to maintain a flow of funding, which not only keeps animals fed but pays the meagre salaries of shelter employees. Most importantly, shelters keep the pet industry from crashing in on itself since shelters control the surplus and thus keep the market for new product healthy.

And there are some within the shelter and rescue chain who stand to make a profit from sustained turnover of "product." I've heard several veterinarians express concern about the interstate (and even international) transport of rescue dogs. Truckloads of dogs are shipped from areas with excess supply—usually the heavy breeding states like Pennsylvania and Missouri—to areas with low supply and high demand, such as the Northeast and the West. These transport schemes undoubtedly save the lives of some dogs. But they also raise ethical concerns. For one thing, interstate transport of dogs is not covered under Animal Welfare Act regulations, so there are no rules about how dogs should be housed and transported, and no regulatory oversight of transport companies. Furthermore, breeders in the southern states see that they have an ongoing market for their dogs. They can easily dump litters that they have been unable to sell, and they can also sometimes sell puppies straight to rescue organizations, who want to fill northern shelters with desirable, adoptable dogs. This, in turn, reinforces the preferential market for purebred dogs and young puppies and decreases the odds of adoption for adolescent or adult shelter dogs, particularly large dogs of uncertain or mixed breed.

Media attention focused briefly on the dog transport issue when an Indiana man who was transporting dogs in his minivan had car trouble. After being tipped off by a motel employee who noticed the dog-filled van in the parking lot, Licking County Police recovered fifty dogs and twelve puppies, some as young as several weeks old. The dogs were packed like sardines, four or five to each small kennel, and had been in the van for at least twenty-four hours before being rescued.[9] There are, at this moment, some unknown number of vans and semitrucks transporting dogs around the country. One person I interviewed said there are "probably at least twenty semitrucks, at any given moment." (He asked to remain off the record.) Pet transport companies like Dog Runner Pet Transport and Peterson Express Transport Services have sprung up to meet the demand for moving large numbers of dogs across the country. Typical "take" for each dog brought up north to an adoption event or breed rescue is about $125 a head. If you have a semitruck and can transport seventy-five to a hundred dogs over the course of a couple of days, that's not a bad income. Is this transporting a problem? In some cases, like the Licking County fiasco, yes. But we also need to be careful not to confuse inhumane transport with humane transport and with the efforts of legitimate rescue groups trying to save the lives of animals who happen to land in a spay-and-neuter challenged part of the country.

What to do? We constantly hear the mantra "adopt a shelter pet!" and surely this is important since these animals may lose their lives if they are not rescued in time. But what if securing an animal from the shelter is also, in other ways, compounding or at least reinforcing some welfare issues that concern us? The most obvious solution, which I mention time and again, is to opt out of the system altogether and not have pets or support any facet of the pet industry. But this is not a solution that the animal lovers among us will want to hear.

Shelter adoptions are still ethically preferable to purchasing an animal on the Internet, at a pet store, or from an unsavory breeder. At present, only about 25 percent of the dogs and cats in people's homes have come from shelters—a woefully low number. But it seems like an uphill battle for shelters. In my town newspaper the *Longmont Daily Times-Call*, the humane society has a section every Sunday for "Pets of the Week," featuring a particular dog or cat whose picture and hope-filled bio ("Charlotte is a sweet girl who is always up for a romp but will just as happily spend the day curled up by your feet") will break your heart every time. Unfortunately, the humane society's pet section has to compete with the classified pet section, which always has at least seven or eight and sometimes more ads for adorable puppies (American Kennel Club registered!) and kitties (purebred Siamese!).

38. Cradle to Grave

One of the lessons taught by people who think carefully about environmental impacts and conscientious consumption is that we ought to consider where the products we buy originated. Are our tomatoes genetically modified, were they grown with pesticides that are killing bird populations and making frogs deformed, were they transported thousands of miles to our grocery store or grown locally, and did immigrant laborers pick the tomatoes for poverty wages? What is the full environmental or moral cost of the product? In the environmental literature, these are called life-cycle or cradle-to-grave analyses. In this case, the "product" of interest is the live animals that are for sale at a place like PetSmart or PetCo. Where did the cute little hamsters or the leopard geckos or hermit crabs come from? What would a cradle-to-grave analysis of a pet look like?

For many of the creatures on the pet-store shelves, their journey has been a living hell. People for the Ethical Treatment of Animals' undercover footage from U.S. Global Exotics, one of the country's largest animal wholesale warehouses, shows animals housed in crowded and filthy conditions. ("Housed" is yet another euphemism common to the pet industry.) Mice and rats and hamsters and guinea pigs are crammed in small bins, where they have to fight for space and for access to food and water. In one piece of video footage, the warehouse has experienced some flooding and a plastic container full of mice is half-submerged and the mice who are still alive are scrambling on top of each other to stay above water.[1]

In another scene, the camera pans across stacks of maybe a hundred or more small tin pie containers with plastic lids. Looking more carefully, you notice that each pie tin contains a coiled snake—the snakes' bodies fill up the small space, leaving them no room to move. The video then pans over to an employee who is holding a plastic two-liter soda bottle upside down, shaking it and pounding the bottom as if to remove something stuck inside. Indeed, what's being forcefully ejected are a number of tiny frogs. Monkeys and hedgehogs are held alone in small cages. Some of them display stereotypic behaviors like pacing—a sign that confinement is driving them mad; others slump against the side of the cage as if they have given up all hope.

Spurred by People for the Ethical Treatment of Animals' undercover work, state authorities in Texas authorized a formal investigation of Global Exotics, carried out by a group of veterinarians, biologists, and others in 2009 and 2010. The entire inventory of nonhuman animals—more than 26,400 animals of 171 species and type—was confiscated. Eighty percent of these animals were found to be "grossly sick, injured, or dead, with the remaining in suspected suboptimal condition."[2] Some of the causes of death included being crushed, dehydration and starvation, thermal stress (conditions either too hot or too cold for a given species), and cannibalism. A court later ruled that all of the animals at the facility—all 26,400—had been treated cruelly. The findings of the investigatory team were subsequently published in the *Journal of Applied Animal Welfare Science*. After detailing the horrors discovered during the seizure, the team writes in academic deadpan, "It is the authors' understanding that these apparently obvious deficiencies reflect normal practices."[3]

Many people already think about where their food comes from. Some people even think about animals, in their purchasing decisions at the grocery store: maybe they don't buy meat at all because of welfare concerns, maybe they only buy meat sourced from certified humane companies, maybe they refuse to buy veal. Why can't this same kind of scrutiny be given to purchases of live animals being sold as pets? It can, but first consumers need to be armed with enough information to start asking the right questions. Consumers and activists alike can push for greater transparency in the pet industry. If we could see behind the scenes of the pet industry, and if breeding and wholesaling facilities had glass walls, I suspect many of us would also lose our appetite for pets.

39. A Living Industry

It ain't about right. It's about the money.

D'Angelo Barksdale (*The Wire*)

The pet industry preys on our love for animals and exploits it. One way the pet industry does this is through cultivating a cultural narrative in which pet keeping is part of a normal and happy life and in which a complete family includes at least one nonhuman member. Within this narrative, pets are loved and cherished and treated with tenderness, just like children. Thus we are told, over and over: nine out of ten pet owners consider their pet a part of the family. Why does this statistic get repeated so often? It is almost like an advertisement for pet keeping. Oh, wait. It *is* an advertisement. Much of the "survey data" that we are fed through advertisements and the media comes straight from industry trade groups such as American Pet Products Association. The media does free advertising for the pet industry by repeating the "data" so often that consumers accept them as fact.

Almost every article about companion animals begins with a statement about how people view pets as family. I admit that I took this sound bite for granted for a long time and repeated it in things that I wrote. What got me to wondering about the pets-are-family claim was researching this book because the extent to which animals are mistreated, abused, and thrown away seems so out of sync with what we are told pet owners are like. Someone who considers himself a parent to his dog is unlikely to bash the dog's head in with a shovel. How can we explain the high levels of neglect, abuse, cruelty, and sexual exploitation of animals, if we are such a pet-loving society?

Let's unpack the pets-as-family claim. The nine-out-of-ten statistic is drawn from a Harris Interactive poll. The data reflect the responses of 2,634 adults to an online survey. The 2,634 people were drawn from among those who had previously agreed to participate in Harris Interactive surveys (which has already narrowed our demographic to Internet users who agree to participate in surveys). From this group, 1,585 reported that they had a pet. This 1,585 is then the base from which the rest of the statistics are drawn. These 1,585 are asked yes/no or Likert-type questions (e.g., "on a scale of 1 to 5, with 5 being the highest"), which would make any student of research methods shake his head in dismay. So the pets-are-family data—if we can call it that—comes from a very, very thin slice of pet owners who are willing to take a survey about pet ownership, based on their answers to a few leading questions. These are just the people who are, I would guess, the type to identify their animal as a family member. These are not the sadists, animal rapists, or even the dog owner whose so-called companion spends his life in the backyard, at the end of a chain—who make up far more than a mere 10 percent of the pet-owing demographic. A gossamer pets-are-family thread has been woven over the ugliness.

Animals are the backbone of an enormous commodity chain. The living creature is the glue that holds an enormous industry together: without live animals, you can't sell cages, tanks, foods, toys, veterinary products and services, or grooming supplies. We would have no need for Snoutstik, or BarkArt blow pens, or PetSweat electrolyte drink for athletic dogs. Nor would there be a market for the billions of pounds of meat-industry waste products. Animals are at the heart, but they are a relatively small organ in the massive body of the pet *industry*. And perhaps counterintuitively, live animals are not where the real money lies.

The family who buys a $10 hamster will spend about $100 on

the initial set up: the Habitrail, the Carefresh wood pulp bedding (in rainbow colors), the Forti-Diet hamster food. And, for as long as the animal lives, this family will spend at least another $20–30 monthly for bedding and food. My daughter's goldfish, Klondike and Dibs, each cost twelve cents. Their twenty-gallon tank cost $90, and with filters and oxygenators and water conditioners and rocks and plants and the little statuette of a treasure chest, we've given substantial support to the pet industry. (All of these things are required of us, as responsible fish owners. And still, I'm afraid Klondike and Dibs have a sad existence, swimming around in circles with nothing much to do.) If we had saved up all the money we spent on pets and pet supplies when my daughter was growing up, we would have had at least a year at an expensive private college already paid for.

So it isn't the cost of animals that really supports the pet industry, but all the supplies you are told you will need. Indeed, according to statistics gathered by a marketing company called Mintel, of the $50.8 billion dollars that consumers spent on pets in 2011, 38 percent was spent on food, 22 percent on supplies and medicine, and 28 percent on vet care. The remaining percentages are left unaccounted for, but I can only surmise that the last 12 percent was spent on the live animals themselves. What does this suggest about the value of companion animals in our culture? Worth about a dime to your dollar, metaphorically speaking. I know that market value is a repulsive accounting of somebody's worth, and many of us would be unable to put a monetary value on the worth of our animal companions, but the pet industry is willing and able. And animals are cheap. Lee Edwards Benning's 1976 book *The Pet Profiteers* called out the industry, and consumers, for what could only be viewed as irresponsible buying habits. We are impetuous and unknowledgeable and spend more time choosing a pair of shoes than a pet. The reason for this may be quite straightforward: we can afford to be impetuous because animals are cheap. We choose our shoes more carefully because they are considerably more expensive.

Pets have something that profitable businesses love: planned obsolescence. Hamsters, gerbils, and rats kept as children's pets rarely live more than a year or two; guinea pigs maybe three or four. Even animals who are relatively long-lived, like goldfish and geckos (who can live up to twenty-five years), rarely survive more than a few years in captivity, if they are lucky. Lots and lots of money is made on the pets who fail (or should we say "die"?). And with a child grieving the loss of a pet, the easiest though perhaps not wisest parental response is to say, "Don't be sad. We'll go pick out another." Or, perhaps the parent will be secretly relieved that the animal has died because the child has gotten the "pet" bug out of his system and the cage or tank and paraphernalia can be sold for $5 at the next yard sale.

The cover story in the February 2014 issue of *Pet Business* is called "What Tomorrow Brings" and explores current challenges that threaten the pet industry's well-being. Three things, we are told, threaten to stall growth: the recession, a shift in demographics, and animal activists. On the shift in demographics, we learn that "industry experts are wringing their hands over what may happen when baby-boomers—noted for their love of pets and their ardent support of the pet industry—decide they have had enough of pet ownership."[1] Will retailers be able to convince a new generation to buy into the pet-owning hobby?

To overcome sticker shock, retailers should promote the wealth of benefits from pet ownership. Retailers and manufacturers need to fine-tune their strategies. "What Tomorrow Brings" bemoans the fact that five-year-olds no longer demand a hamster or a bunny but instead want a game console. "We are really dedicated to getting people back into the hobby" of pet owning.[2] To accomplish this, *Pet Business* recommends targeting millennials, with particular focus on niche markets such as Hispanic pet owners, one of the few segments of the market that is showing growth: "Hispanic pet owners are willing to spend less on vet care, but more on services and discretionary extras. . . . [They] are more likely to own pet clothing and are more interested in treating their pets to cosmetic luxuries."[3] How's that for racial profiling?

In the same issue of *Pet Business*, a section called "A Living Industry" focuses on the most pressing threat faced by the industry, which "lies at its very core": the availability of pets. The animal activists are of particular concern in this regard because, even though they are "well-intentioned," they "might just succeed in knocking out the foundation upon which the industry was built." How? By persuading the public, and lawmakers, that the sale of live animals must be more tightly regulated. "If ambitious animal activists succeed, will there be any pets available at all?"[4] There is tremendous movement to limit the sale of live animals, says Michael Canning, president of the Pet Industry Joint Advisory Council, or PIJAC, a pet industry lobbying group (see chap. 40, "Protect the Harvest"; also chap. 46, "Offering Better Protection"). And the CEO of Petland, Inc., warns the readers of *Pet Business*, "People—and pet retailers—are losing their rights to make responsible decisions on the types of pets they can bring into their homes and where they choose to acquire those pets."[5]

The animal activists have mostly focused on dogs, but you can see them starting to broaden their attack to all segments of the pet trade, to include cats, reptiles, birds, even small mammals. The attacks are based on "matters of animal cruelty" and "questionable sustainability when it comes to procurement." Canning goes on to say: "Those of us at PIJAC, responsible pet manufacturers and distributors see the real need for shelters . . . but there is a consequence to having animals that are spayed and neutered become the majority of pets out there." The "amount of breeding dogs" could decline precipitously and the industry must not "stand idly by." "The challenge for the pet industry is to educate the American public as to the benefits of sharing our lives with pets."[6]

To convince us of the benefits of sharing our lives with pets, the industry spins the narrative of pets as a happy and necessary part of every healthy family. Not only that, but the industry convinces us that we can all engage in "responsible pet ownership." In this way, the industry can quell any moral hesitations and offset suspicions that animals themselves don't benefit, and in fact suffer great harm, because of our obsession with pet keeping.

40. Protect the Harvest

I find it odd that pet industry lobbyists spend their time trying to fight regulations and restrictions aimed at protecting pets. But I suppose this makes sense: protections for animals would likely make selling them more difficult and less profitable. And the pet industry must sell pets and pet keeping—there is a great deal at stake. In 2012, annual expenditures by U.S. consumers on pets and pet products hit over $55 billion dollars, reflecting steady and strong growth over the past two decades (spending in 1994 was only $17 billion; the industry has never had a down year since then). As we've seen, what drives the pet industry is sustainability, in the corporate sense. How can the pet industry continue to flourish and even grow? By selling more animals and thus, also, more animal-related goods and services.

Perhaps the most notable pet industry lobbying outfit is PIJAC or the Pet Industry Joint Advisory Council, whom we met in the last chapter. The council is dedicated to keeping the pet industry economically healthy, and one of the biggest threats to the sale of pets is attempts to legally regulate how pet animals can be treated. So PIJAC is busy lobbying to defeat rules that would ban the retail sale of dogs and cats and close the loophole that allows for the unfettered sale of live animals over the Internet. I noticed that PIJAC also has a Marine Ornamental Defense Fund, which is trying to protect fish hobbyists from legislation that would prohibit the importation of endangered coral and endangered fish such as the clown fish and damselfish.[1] They ask you to please donate, so that your *right* to be a fish hobbyist is protected.

This is an interesting use of language, and here is what I see happening: there is a great deal of manipulating of consumer attitudes and preferences. Once these core habits are instilled, they are then called our "rights." So the narrative goes even deeper than happy pets making their families happy and healthy. Indeed, narrative becomes ideology: owning animals is part of our American heritage of independence and freedom.

Here I insert, as Exhibit A, the National Animal Interest Alliance's "Dog ownership . . . a cherished American Tradition" campaign. "George Washington had one, and so did Benjamin Franklin. But if animal extremists have their way, your grandkids won't."[2] I appreciate that the National Animal Interest Alliance encourages responsible animal husbandry, yet their insistence that owning animals is our right as American citizens and that legal restrictions on what people can do with and to animals is an infringement of our liberties strikes me as overstated, to say the least.

What happens with PIJAC and the National Animal Interest Alliance and other such industry groups is that the rights of animal owners are strategically placed in tension with the rights of animals, with "animal rights" being a crude caricature of an otherwise extremely nuanced and complex line of argument explored by moral philosophers. Consider, as Exhibit B, the advocacy material for Protect the Harvest.[3] This political advocacy group is dedicated to protecting our God-given rights to hunt, fish, farm, and, as if that weren't enough, to own pets. Harvest is particularly concerned about the activities of a "radical animal rights" organization called the Humane Society of the United States. The HSUS, warns Harvest, are extremists who believe that animals have exactly the same rights as humans. Take, for example, the HSUS's support of initiatives to ban the retail sale of dogs and cats in pet stores. The HSUS people, says Harvest, "are not just attacking the breeder's ability to provide for themselves [*sic*] and their families, but they are targeting your right to raise and care for your family's pets."

These oversimplifications and linguistic deceptions will offend the intelligence of anyone who has given even a little thought to the moral intricacies of human relationships with animals. You can be opposed to some forms of hunting (e.g., "canned hunts," where the animals are essentially captive and are lured, by food or salt licks, straight into a hunter's sights) or some forms of fishing (e.g., the use of barbed hooks) without condemning the entire enterprise. So, too, is pet keeping subject to a great deal of moral nuance. Framing any attempt to protect companion animals as "animal rights" and placing this in opposition to human rights is as illogical as it is underhanded. And it creates an impossible situation for animals.

41. Rent-a-Pet

One of the latest trends in the pet business is to lease animals, rather than sell them. Hannah the Pet Society, which opened its first two retail locations in 2010, is the first business venture into this new frontier of pet keeping. The rent-a-pet concept is the brainchild of Scott Campbell, the veterinarian who built the Banfield Pet Hospitals empire, the largest chain of vet clinics anywhere in the world. If you've been inside a PetSmart, you've seen a Banfield, which is why Banfield has the nickname "vet in a box." The business strategy of Banfield has been to standardize veterinary care and offer a low-cost package that everyone can afford—and it is super easy when you can buy your animals, their food and toys, and their veterinary care all in the same place. The master plan is to expand all over the United States and eventually, like Walmart, become a global presence. In 2007, Campbell sold Banfield to Mars, Incorporated, the maker of M&Ms and Snickers and also several popular brands of dog food, such as Nutro, Pedigree, and our old friend Cesar and his Canine Cuisine (see chap. 17, "Feeding Frenzy"). Yearly revenue for Banfield is $187.5 million.

"Hannah is here to help you overcome common Pet care challenges like behavioral concerns, unexpected veterinary expenses, the cost of Pet food/supplies and unfavorable past experiences."[1] Thank goodness someone has finally offered to take all the hard work out of Pet ownership—or, as it is called at Hannah, Pet Parenting. (At Hannah, all animals are referred to as Pets, with a capital *P*.) Campbell's pet-leasing business is named after his mother, Hannah, who loved animals. Although the "Society" in the name makes it sound like a humane society of some sort, don't be confused: Hannah is a for-profit pet-selling business.

So how does it work? The basic idea is that Hannah will find you the perfect Pet; you sign a contract to pay for the Pet as well as the Pet's Lifetime Wellness Plan. (You must get all veterinary care at one of the two Hannah clinics.) If you are not happy with your Pet, you can come in and exchange the Pet for a different one.

To help you select your Pet, you will use Hannah's proprietary Lifetime Matching Program. Created with the help of "psychologists, veterinary behaviorists and personality testing experts," this computerized matching software (called Hannahware) is sort of like interspecies Internet dating, except we get to ask all the questions. The Pet Matching team at Hannah will give you a questionnaire that elicits important information about your personality, the particular qualities you would like to find in a Pet, how much experience you have and how much time you want to spend, and then Hannahware will churn out a perfect match. Not only will it help you decide whether you should get a guinea pig or a dog, but the Pet Matching team will also search their database of available animals to find you the perfect individual. (Your Pet options at Hannah are limited to dogs, cats, rabbits, and guinea pigs.)

Once Hannah has secured the Perfect Pet, you then sign a leasing contract. Along with your Pet, the contract also provides you Total Lifetime Care. The veterinary needs of your Pet are supposedly taken care of: preventive care, emergency care, grooming, and nail clipping (all at a Hannah location). You can even buy a premium plan, which includes having Pet food delivered to your door once a month and having your Pet picked up by the Hannah Limo and taken to and from grooming or pet appointments or to training classes. Really, as I said before, all the fuss of Pet ownership is removed.

Now, the guarantee part: if, for any reason, you do not like the Pet with whom Hannah has matched you, they promise to retrieve and rehome the Pet and find you a more suitable one. Hannah customers sign an Enrollment Agreement—essentially a lease—which initially covers a five-month-long Honeymoon Period, during which time the Pet Parent can decide whether or not the Pet is a good match. At the end of this Honeymoon, the Parent has the option to continue leasing or to purchase the animal at the previously agreed-on price, as stated in the contract.

My initial reaction to Hannah was revulsion: How can you even talk about "renting" a living being? But I've tried to have an open mind and consider the positives. People often don't really know what kind of Pet is best for them—what breed, or even what species. Usually if you buy a Pet and discover after a few weeks that you don't, for example, really like the smell of guinea pigs, then you are stuck with your bad choice. And what can you do, other than just stick it out and resent the animal, pawn her off on a friend, or take her to the shelter—all of which take time and effort and make you feel guilty? Hannah removes the risk. Hannah helps people think through what they want from a Pet, and what kind of Pet will best fulfill their expectations and desires.

Another good thing about Hannah is that since veterinary care is part of the package, Pet Parents might be more likely to make good use of the services, not having to weigh what might seem like a minor veterinary problem against whether the vet bill is "worth it." Hannah also provides educational and training support, so that a family has a ready resource for working through behavioral challenges, which increases the chances of a successful relationship. Although Hannah's prices might seem rather high, they actually aren't too bad (like Walmart, the business model is one of high volume and low prices) and they can help regularize cost: you can plan to spend a set amount each month, regardless of veterinary emergencies and such. This may make having a Pet more viable for people living on a fixed budget. Finally, Hannah's system might provide peace of mind to an elderly person who wants to share her home with a Pet, but who worries that her animal might outlive her—because the Pet will go back to Hannah and be rehomed.

Where do the animals come from? This is one of key selling points of Hannah—and also the primary source of criticism. Hannah assures its customers that all animals come from area shelters, rescue groups, foster homes, and families who have sold their unwanted animal to Hannah. But animal welfare groups have questioned whether Hannah hasn't oversold its role as a "rescue" organization, particularly when Hannah regularly provides people with designer pups like labradoodles and shih-poos, which are rarely found in shelter settings. As we've already seen (see chap. 37. "The Shelter Industry"), there is no set definition for "animal rescue," nor is sourcing animals from a rescue any guarantee that they didn't begin life in a puppy mill. Many of the Oregon and Washington shelters near the Hannah stores refuse to provide Hannah with animals, and one has to wonder why.

Perhaps the most unsettling thing about Hannah is its stated goal: "Pet Parenthood. Redefined." We might celebrate that humans are no longer called "owners" (unless they choose to buy after the Honeymoon) but is "renter" really a better model for our relationship to companion animals?

Putting aside the dubious prospect of pet rental, there may be creative ways to bring people and animals together that allow humans who crave contact with an animal freedom from responsibility. For example, a new trend—popular in Japan and catching on in other parts of the world, including the United States—is the so-called cat café. At Meow Parlour in New York, customers can buy cat-themed pastries at the café and, right next door, visit with cats. Time slots can be reserved online ($4 for each half hour). You can sit and work on your computer, surrounded by cats. Or you can go into the playroom and interact with cats. There are rules that ensure the cats have their personal space respected. All of the cats technically belong to a local shelter and are available for adoption. The Meow Parlour provides prospective adopters as much time as they want to interact with a group of cats and see if they feel a particular connection with one special cat. It also allows cat lovers who have allergic spouses or who live in no-pet apartments or who work long hours to get their "cat fix."[2] Another innovative way to connect people and animals, without long-term commitment, is a phone app called walkzee. Walkzee allows people who are traveling to "borrow" a local shelter dog for a few hours and take him or her for a hike or walk. As long as "pet-borrowing" schemes such as these are aimed primarily at improving the lives of the animals themselves, they seem more ethically sound than the rent-a-pet model.

42. The Biggest Loser: Exotic Pets

A prisoner's got to have some kind of a dumb pet,
and if a rattlesnake hain't ever been tried, why, there's more
glory to be gained in your being the first ever to try it.

Mark Twain, *The Adventures of Huckleberry Finn*

The documentary movie *The Elephant in the Living Room* is a look inside one of the more bizarre subcultures in America: the world of exotic and dangerous pet ownership. Most of the film takes place in Ohio, which seems to be the epicenter of exotic pet culture. We follow the interwoven stories of Tim Harrison, an Ohio public safety officer who works for an organization called Outreach for Animals, and a man named Terry, who keeps several lions as pets in his backyard. Tim spends his days trying to track down and catch the various exotic pets that for one reason or another end up on the loose in the neighborhoods around Dayton. Terry spends his days struggling to care for his outsized pets, which becomes especially difficult after his large male escapes from his pen and starts chasing down and attacking cars on the highway.

It is hard to watch the footage of Ohio's exotic pet auctions and reptile expos and not be disturbed by animals being traded like baseball cards and venomous snakes being sold to children as toys. Still, the overriding tone of the movie is not outrage, but sadness. There really are no winners. For Terry, things end in heartbreak.

He loses his beloved male lion, Lambert, to a freak electrocution, and finally realizes that he can no longer shoulder the burden of caring for the female, Lacy, and her three cubs. For Tim, things also end in heartbreak—day after day as he tries to deal with an increasing number of abandoned, lost, or escaped animals. Most of all, we feel sad for the animals, for whom things always seem to go badly. As Tim says at one point, "There are no happy endings for the animals."

There is no precise definition of "exotic pet." By some definitions, an "exotic" is any pet other than a dog, cat, horse, or fish. So, sugar gliders, geckos, and hermit crabs—though common pets in U.S. households—are exotic. "Exotic" can also refer to any nonnative animal (a lion in Ohio) or any nondomesticated animal (a garter snake). We don't need to pin down a definition in order to understand the problem, but it is important to note that lack of clear definition makes trying to regulate, or even respond carefully to the issue, more challenging. The American legal system doesn't even have consensus on how to define "animal" so laws protecting such entities are necessarily disorganized and confusing and thus difficult to enforce. A glance through state laws on possessing, selling, importing, and exporting "exotics" reveals wide diversity in definition, as well as regulation.

Various things about exotics make their care challenging. Most people know nothing about the behavior and natural history of the bearded dragon or capybara, and providing appropriate care requires significant effort to understand the animal and his or her specific needs. Not everyone is willing to put in this effort, and often the animals are made to live in conditions that are unhealthy. Exotic species are not adapted to our human habitats, as are dogs and cats, and have specialized living requirements that are often hard to mimic (e.g., in the wild, a male lion might have a territorial range of up to forty square miles). Depending on where you live, it can be hard to find a veterinarian who is trained to care for exotics, and the more exotic the creature, the more difficult this becomes. Many exotics are kept illegally, decreasing the chances that an animal will receive care from a trained professional.

According to a recent study by Emma R. Bush and colleagues in *Conservation Biology*, "Animal welfare is compromised to some extent at all stages of the exotic pet trade."[1] The first challenge for an animal coming directly from the wild is capture. Bird capture methods, according to this study, "include painting a sticky resin to trees that damages feathers and limbs." When primates are the quarry, a common capture method is simply to kill the adult and take her young. Once animals are taken from their wild homes and families, their trials have really just begun. They have to adjust to radical changes in their behavior, diet, and environment. Wild birds, for example, must be acclimated to the dry bird food available to pet owners, which is often unsuitable for the species or age of bird.[2] There is, according to Bush et al., a scale of suitability to captive life, depending on "the temperament of the animal and the complexity of its needs that is linked to its level of specialization in the wild." Species at the extreme end of the unsuitable spectrum are referred to as "cut flowers." For example: "The slow loris (*Nycticebus* sp.) is sold at a relatively low price because it does not survive long in captivity following the removal of its teeth to prevent a toxic bite."[3]

Millions of exotic animals are held in captivity by private individuals, all around the country. For example, there are over fifteen thousand big cats owned by people in the United States. There are more tigers in captivity in Texas than there are in the wild. I could buy a tiger for less than I would pay for a purebred dog or cat, and I could treat it however I like. There are no regulations, or even common standards, for how these animals ought to be cared for, and complete ignorance about the behavior and natural history of a creature is no impediment to buying one.

The animals people choose for pets is quite astonishing, really, particularly when there are children also present in the home. Crocodiles, venomous snakes, lions, bears, wolves, chimpanzees. It may be that the dangerousness of these animals is what draws people to them. And I suppose we could just say caveat emptor to Joe, who decides it would be fun to buy a baby Burmese python and name it Monty. We could just say, "Joe, if you get squeezed to death, you asked for it." The problem is that Joe's choice has an impact on more than just Joe. Most obviously, there are implications for the Monty the python, who is held in captivity, who has no hope for a "normal" python existence. Monty may receive substandard care, may die after a few months in captivity, or if he is lucky enough to survive, may eventually be abandoned when he gets too big. The purchase of Monty supports and encourages the live animal trade. There are impacts to the environment such as the large number of Burmese pythons who now inhabit the Florida Everglades and are killing other animals. Finally, Burmese pythons can be dangerous to people.

Exotic pet keeping goes beyond the so-called megafauna—the glorious tigers and giraffes and lemurs. There is also a fascination with smaller creatures, and the stranger the better. This fascination with the bizarre is often glorified by the media. Animal Planet, for example, ran a show called the "Top Ten Peculiar Pets." Among the animals featured on this show were the wallaby, the capybara, stick insects, Madagascar hissing cockroaches, and mini donkeys. These creatures are surprisingly easy to buy. Oddballpetfactory .com, for example, tells us to "Come in and see our hand picked, captive bred little charmers." Oddball carries almost all of the critters on Animal Planet's list, plus some other interesting choices (worms, black widows, raccoons). In Japan, which has a long tradition of mushi or pet insects, you can now buy living creatures from a vending machine. Although a few exotics might do fine in captivity (the mini donkeys and the potbellied pigs, for instance), many of these exotic "pets" are really just trinkets.

What about the animals themselves? Do they benefit in any way? One of the key things about exotics is that these species have never been domesticated. Whether they were born in the wild or bred in some warehouse in Ohio and sold at auction is beside the point: they are still wild. People claim to bond with these animals—and perhaps they do, though it is hard to know if the animals experience any warm feelings in return. Tolerance and indifference can be mistaken for affection.

Could we envision any benefits to keeping exotic animals as

pets? It might be an interesting and exciting adventure to interact with an animal who, unlike a dog or a cat, really seems like an "Other" and retains his or her wildness. Perhaps this experience might connect us with nature in a way that we have largely lost, now that we live in concrete cities and travel by car and eat from Styrofoam packages. It might make us fascinating to our friends and to potential mates. We might have a strong attraction toward some particular kind of animal, like lions or elephants or pythons, and want to commune with them. We might think our children will learn cool things and grow up to be biologists if we supply them with interesting animals to pique their interest. But these are all selfish reasons for keeping an exotic, none of them particularly compelling, and none of them worth the suffering and heavy loss of life for these animals.

Some people argue that wild animals kept as pets have cushy lives because they have ready-made shelter and don't have to hunt for their own food. But the cushy life a human keeper provides is basically a prison. And what looks "cushy" to us might be unsuitable in the extreme for the animal. Animals are denied nearly all of their natural behaviors, not to mention their freedom, and it would be hard to argue that their lives aren't severely diminished by captivity.

The Pet Industry Joint Advisory Council maintains that individual hobbyists aid in conservation efforts because they learn, through personal experience, how best to care for a particular species of animal—and this knowledge could be shared with others. I seriously doubt that conservationists would find this a compelling line of thought. And then there are those (we've met them already) who claim that regardless of the welfare of animals, owning an exotic is a matter of constitutional rights. To which I say: these folks need to read the Constitution a little more carefully.

The Biggest Loser in the exotic pet industry? The animals.

Caring for Spot

43. What Do Pets Need?

When we bring an animal into our home as a pet, the quality of this creature's life is utterly in our hands. Unlike our children, who are an awesome responsibility but will eventually grow up to become independent beings, our animals will rely on us from birth to death and everything in between. Their lives are inextricably bound with ours, and it is my humble opinion that once we have chosen to bring an animal into our home, we have an obligation to do the best we can to provide this animal with a good life.

Animals need unconditional access to clean water, nutritious food, shelter from the elements, and adequate amounts of exercise. They also need frequent enough veterinary care that treatable illnesses are dealt with and painful conditions either resolved through treatment or addressed with adequate medications. Even something as seemingly simple as having a seasonal or food allergy can cause significant distress, and it is unfair to leave such conditions untreated. (We've all been itchy at one point or another and know just how awful it can be.) As with human parenting, there will always be disagreement about what optimal physical care looks like and where to draw the lines among excellent and adequate and negligent care. Is an animal (or child) who is allowed to become slightly overweight being mistreated or overloved? Is two hours of exercise a day "adequate"? These sorts of questions can't really be answered in the abstract (your dog might need a lot less exercise than mine), and perhaps they can't be answered with certainty at all. Ultimately, common sense and compassion can guide people to do the best they can.

What do animals need to be emotionally healthy? The answer, of course, depends on the species of animal and, perhaps even more, the individual animal him or herself. As a basic rule of thumb, the intention should be to provide an animal with a life in which positive, pleasurable experiences far outweigh negative, unpleasant experiences. In other words, provide them good quality of life. Well-being isn't something you have or don't have; you have it to some degree or another. Negative feelings—pain, fear, frustration, itchiness—are an inevitable fact of life, but we can work hard (as we do for children) to keep the balance as far to the positive as possible. We can help foster a feeling of security by providing an environment sensitive to the sensory experiences of our animal, including his or her sense of smell, and sensitivity to various kinds of lighting and noise. We can provide our animal opportunities to make behavioral choices such as to hide or play; the option of saying no (e.g., a dog should be able to decline unwanted petting from a stranger); opportunities for species-specific behaviors (e.g., dust baths, scratching stuff for cats); and an environment that increases behavioral diversity. And we can provide our animal with ample social companionship, both with us and with others of their own kind.

One of the most important aspects of providing good care is making sure that an animal's needs are being met consistently and predictably. Like humans, animals need a sense of control. So an animal who may get enough food but doesn't know when the food will appear and can discern no consistent schedule may experience distress. We can provide a sense of control by ensuring that our animal's environment is predictable: there is always water available and always in the same place. There is always food when we get up in the morning and after our evening walk. There will always be a time and place to eliminate, without having to hold things in to the point of discomfort. Human companions can display consistent emotional support, rather than providing love one moment and withholding affection the next. When animals know what to expect, they can feel more confident and calm.

A common misconception is that pets have easy lives. They don't have to do any work to find food and shelter or to protect themselves from harm. But making life easy for captive animals doesn't do them the great favor we might imagine. Providing them with appropriate challenges affords them opportunities to put their functional competencies to work, to engage in their full range of behaviors, and to engage their intelligence.[1] And, in fact, various studies show that animals *like* to work and will engage in work for a reward, even if the reward is otherwise available for free.

"Agency" has recently entered the vocabulary of animal welfare science and captures an important element of what animals in captivity need. According to Marek Špinka and Françoise Wemelsfelder, "agency" can be defined as "the propensity of an animal to engage actively with the environment with the main purpose of gathering knowledge and enhancing its skills for future use."[2] It is "an integrative capacity that works across specific modules of organization and, as such, forms an important condition for an animal's overall health and well-being."[3] Animals, they argue, have an intrinsic tendency to behave "beyond the degree dictated by momentary needs" and to continually broaden their range of competencies. Prominent facets of agency include problem solving, exploration, and play—each of which provides the opportunity to broaden competencies and each of which is an intrinsically rewarding activity. An animal's natural environment is complex and constantly changing; it is an open world, in which she can always expand the horizon of her knowledge. The captive environment, by contrast, tends to be barren and unchallenging. In order to engage an animal's agency, captive environments need to be rich and complex and provide opportunities for problem solving, exploration, and play.

44. Enriching Animals' Lives

There is ample empirical evidence, dating back at least four decades, that environmental enrichment does improve animal welfare. But we need to have a handle on the details of exactly what provides enrichment for which animals, how much, and why, and in this regard we still have much work to do. Animal behavior professor Robert Young, in his book *Environmental Enrichment for Captive Animals*, suggests that we know much more about how to enrich the lives of animals in zoos and labs than those in home environments. Most species kept as pets are also routinely used in other settings. For example, dogs are used in toxicology studies; cats are used to test the effectiveness of military weapons; rats, mice, guinea pigs, ferrets, rabbits are used in medical research and product testing. Exotic birds and reptiles are housed in zoos. So we have a lot of information about captive housing requirements and other aspects of welfare. Unfortunately, though, "we have virtually no information on the welfare of these species within the home environment."[1] Indeed, we have traditionally only thought about enrichment of pets' environments in what you might call a reactive mode: when they begin displaying behavior that is either worrisome (feather plucking, obsessive licking) or destructive.

Why hasn't there been a more aggressive investigation of enrichment for pets? "It has been suggested to me by various scientists," says Young, "that the topic is too controversial and emotionally charged to touch because we are often talking about a 'loved family member.'"[2] To question companion animal welfare would come across as criticism of owners. It might also come across as a condemnation of the entire pet-keeping enterprise (this blasphemy is mine, not Young's). Still, there are welfare issues that obviously need to be addressed. Why else would there be thousands of books on behavioral problems of dogs and cats and Vietnamese potbellied pigs, a growing army of behavioral consultants and veterinary behaviorists, and an entire genre of reality TV shows dedicated to crazed animals and their equally crazed owners? The titles of these shows—*My Cat from Hell*; *It's Me or the Dog*—suggest something of the desperation pet owners and animals are feeling. "All this," says Young, "suggests that the level of welfare problems for companion animals is not insignificant. . . . Clearly, . . . 'love' is not enough to ensure that a pet experiences a good level of welfare."[3]

"Enrichment," as the term suggests, gives attention to how the environments of captive animals can be designed to enhance an animal's role in organizing his or her own life.[4] For example, we can give an animal opportunities to forage for food, build a social group, or construct his own shelter. It won't work to simply turn on the Dog TV channel. Even though Dog TV claims to keep your canine entertained by watching videos of squirrels running up trees, all it does is provide sensory "noise." Likewise, a running wheel in a hamster cage offers little more than a physical outlet for frustration—better than nothing, perhaps. But not good enough. As Wemelsfelder explains, "To be able to create a meaningful life, the animal must be provided with materials that are biologically salient and enable it to fulfill its primary needs in an inventive, varying, and flexibly adaptive way."[5]

Most companion animals receive two types of enrichment: social enrichment and object enrichment. For dogs, human contact is one of the most enriching things we can offer. We can also offer the opportunity to play and interact with other dogs. As most cat owners know, there is an art to knowing when the fickle feline will appreciate a scratch on the forehead and when she would rather not have our dirty fingers touching her fine coat. And for some species, as Young points out, "human handling is completely inappropriate because the species perceives the human as a predator."[6] With reptiles and pocket pets in particular, people sometimes confuse what appears to be "calm" behavior with what is in fact an antipredatory response of freezing or feigning death. A caretaker may think he is enriching an animal's life when actually all he is doing is stressing the animal. Simply because an animal doesn't avoid human contact does not mean that he or she actually enjoys it, and making sure that handling is actually a positive experience for the animal requires being thoroughly educated about a particular species and its behavior.

Object enrichment can take the form of toys, puzzles, Kongs full of frozen peanut butter, chewable hideouts for rats or hamsters, and all manner of other entertainments. Toys can be great, but often what is most enriching about the toy is not the object itself but the chance to interact with human and toy together, particularly for dogs. Case in point: because Bella loves chasing balls so much, we decided to get her a GoDogGo! automatic ball launcher. We imagined hours of pleasant relaxation in a hammock while she entertained herself. All you have to do is teach your dog to drop a tennis ball into a bucket, and the machine spits out a flying ball, over and over. Bella doesn't use it, though she is plenty smart enough to figure out what needs to be done. She wants *us* to throw the ball because ball chase is, obviously, a two-person game.

It is also useful to keep in mind that not all games and toys and other object enrichments are actually pleasurable for an animal. I recently came across an article about how laser lights aren't good toys for cats, if they are used in the typical fashion: make the little red light dance across the floor, and when the cat almost gets to it, make it go somewhere else. This is basically an exercise in frustration for a cat. Every time he thinks he finally has his "prey," it disappears. Cats need a chance to actually catch their prey now and then. Also, lasers can be a bit dangerous. If they are shone straight into the eye, they can cause injury.

Enrichment isn't just for dogs and cats. One of the main problems facing small mammals, reptiles, amphibians, birds, and fish is that many—I would venture to say most—people who acquire these animals as pets for themselves or their children do so because they are easy and cheap and don't require much care. The goldfish is the classic example of low maintenance. All you do is buy a little bowl, put the fish in, and voilà: you have a pet! But what a sad life for the little goldfish. There is growing recognition that fish are cognitively complex creatures and that captive environments can be far more attentive to their cognitive needs.[7] Fish need mental stimulation, variety, the opportunity to socialize, and the chance to learn. Some interesting research suggests that captive fish kept in enriched environments have improved neural plasticity and spatial learning skills, compared to fish held in a traditional boring tank.[8] Enrichment of the environment can be achieved by altering periods of light and dark or adding novel elements to the tank, such as plants, rocks, logs, and other objects that add spatial variety (like those buried-treasure tank ornaments). Dietary enrichment can include providing a novel food or feeding at different times, and we can provide social enrichment through same-species companionship. We can also provide mental stimulation through behavioral training, such as by clicker training a goldfish (using a flashlight blink for the "click") to swim through a hoop.[9]

People can get marvelously creative in providing enrichment for their animals. I've seen elaborate mazes created for pet rats; tortoises who play fetch with a ball; rabbit playgrounds the size of people's basements, full of obstacles, hidden treasures, and things to explore. I recently watched a YouTube video of a delightful pair of guinea pigs who had been taught to play basketball.[10] In our efforts to engage the creativity of our animal friends, our own creativity is our best resource.

45. Which Animals Should Be Pets?

If we were going to try to take the animal's point of view, which creatures would likely choose to share their lives with humans in our homes and pet stores, and which ones would say "no, thank you" and scamper or slither or swim away as fast as they could? If we opened the cages and doors, I suspect we would look around and find the pet-store shelves quite bare. Dogs would probably stick around, as might cats. But otherwise, I think we would see a mass exodus.

To phrase the question in human ethical terms, which animals can be kept humanely as companions and which cannot? The answer depends on two things. First, what particular burdens do captivity and confinement place on a given species? And second, what are the possibilities for reciprocal and mutually fulfilling companionable relationships with humans? The fewer the burdens and greater the capacity for reciprocity, the more ethically sound the pet keeping. Using these criteria, I would rank species in the following rough order, from most ethically acceptable to least: dogs, cats, rabbits, small domesticated mammals (rats, hamsters, gerbils), fish, birds, reptiles and amphibians, insects, exotics.

Dogs and cats can live well with humans, can be provided with substantial physical freedom and social stimulation, can be truly integrated into human families, and can even be gainfully employed doing work they enjoy. These "über-domesticated" species have lived in close collaboration with humans for thousands of years and have been domesticated as partners and companions to humans, not usually or only as a source of food or fur or labor. Although I think cats can live happily with humans, permanent captivity indoors is problematic, in my view (though some cat behaviorists disagree).

The domesticated varieties of the small mammals can be reasonably comfortable in human company and can form bonds of friendship, though perhaps not at the level of intimacy attained by cats and dogs. However, providing them the conditions they need to be happy, and not just free of physical distress, requires knowledge, commitment, and time. Consider, for instance, two different rabbits. Rabbit Number One was given to a child for Easter. She lives alone in a hutch, out in the backyard. Rabbit One may get taken out and played with most days but otherwise she just sits in her hutch eating Kaytee rabbit pellets. Rabbit Number Two lives with someone who is dedicated to rabbits, knows a lot about rabbit behavior, and has read about rabbit keeping on the House Rabbit Society website. Rabbit Number Two, who lives with several other rabbit companions, enjoys the run of an entire basement and also gets to be in a grassy backyard on nice days. She has a cage she can go into when she wants to feel safe, but the door is always open. She has lots of space to move around, has interesting things to explore, enjoys a variety of different fresh vegetables, and spends several hours each day with her human friend. It is also possible to keep rats, mice, hamsters, and gerbils under conditions in which they can flourish, like Rabbit Two. But it is equally possible—and unfortunately much easier—to do things poorly, and most small creatures live under conditions more like Rabbit One's.

For reptiles and amphibians, captivity carries heavy costs. There may be herpetologists out there with the know-how to provide for these creatures such that there is little physical distress or suffering, but the majority of pet owners will be unprepared for the challenges these animals present. Whether reptiles can truly flourish when their entire lives are spent within four glass walls is a question I leave up to those who study reptile cognition and behavior. But Gordon Burghardt, a highly regarded research ethologist, believes that no captive environment can be adequately stimulating for reptiles and amphibians, even for small or sedentary species, and that the best we can hope for is "controlled deprivation."[1]

Captivity is also hard on birds and fish. Although it is possible to provide them with large and relatively stimulating environments, we cannot compete with the open sky or an ocean habitat. As for insects, we may assume that they have little in the way of "rich experiences" and thus aren't the kind of creature that can be harmed by captivity or early death. But the more we are learning about insects, the more we understand that they have their own complex forms of insect cognition, engage in social relationships with others of their kind, and are exquisitely suited to their unique environments. Our compassion can extend to these forms of life. We have little to gain from keeping insects as caged pets—they are entertaining, but they do not interact with or form social bonds with us—and they have a lot to lose. Exotics just shouldn't be kept as pets, period.

The only pets that really don't need much in the way of care or enrichment, and who can be held captive without any moral concern, are Chia pets (which still require regular watering) and pet rocks. These are great pet choices for the person with a passing fancy or the parent whose child longs for a pet but who doesn't want any responsibility or the person who travels all the time. (From one seller of pet rocks: "You'll find a loyal friend in the USB Pet Rock. It needs no water or food, doesn't make a mess of your home, and yet will always love you in its silent rocky way.") Another good substitute, especially for children (but adults love them, too) are electronic pets, like Furby and Aquabot goldfish, digital pets like Tamagotchi, and virtual companions like Foo-Pets ("Adopt them. Breed them. Love them.")[2] You may think I'm joking, but I'm quite serious. These toys satisfy a range of human pet-keeping impulses, without the collateral damage to real animals. When an "owner" loses interest, nobody gets hurt. Some might object that promoting these unreal pets is a bad idea because it reinforces the idea that pets are toys. But it could go the other way, too, and reinforce the notion that real animals have lives of their own, have complex needs, and shouldn't be sold as objects for our amusement.

46. Offering Better Protection

It is striking that even the most basic rights thought to extend to all persons—the right not to be arbitrarily killed, the right to be free from torture and slavery, and the right not to be arbitrarily punished—are systematically denied to companion animals. I am hoping a time will come when humans are not the only animal to have meaningful legal protections. But we will make little headway in protecting animals if we can't build consensus on their moral value, and this consensus seems still a long way off. For example, according to data collected recently by Faunalytics (formerly the Humane Research Council), only 35 percent of people strongly support the goal to "minimize and eventually eliminate all forms of animal cruelty and suffering." Only about half of the people surveyed agree that humans have an obligation to avoid harm to animals in general.[1] As with our pet industry surveys, Faunalytics data surely capture only a tiny segment of the population, and we can only hope that the results are skewed. These are distressing numbers (though they could be worse) and suggest that animals will remain vulnerable to human excesses at least in the short run.

The rights of pet owners are, for the time being, well-protected. Are there ways, within our current milieu, to also improve the protections offered to companion animals themselves? All of the following have been proposed as incremental changes that would offer increased protections.

licensing requirements for all pet owners
laws limiting or prohibiting the sale of live animals
laws regulating international and interstate shipping of live animals
a federal prohibition on the sale of crush films, in particular, and animal pornography in general
state laws making sexual assault of an animal punishable (not limited to sexual assaults that are fatal or cause severe injury)
better and more frequent inspections of breeding facilities
better and more frequent inspections of animal wholesale facilities
greater transparency in the pet industry, such as, perhaps, in identifying the sourcing of animals for sale
greater transparency in the shelter industry
state laws requiring at least eight hours of training for anyone performing euthanasia
free speech protections for those who expose corporate animal abuses
reporting requirements for veterinarians (e.g., abuse, sexual assault)
combined/coordinated reporting of animal abuse and domestic partner, child, and elder abuse
a publicly accessible national registry of those convicted of animal cruelty or sexual assault
increased (and responsible) media reporting of crimes against animals
more community resources (e.g., tax money) dedicated to shelters, animal control facilities, and cruelty investigators
state-appointed lawyers to represent animals in court
required humane education in schools
laws making failure to provide timely veterinary care a legally enforceable welfare violation
laws allowing pet owners to collect damages for emotional pain

and suffering resulting from the loss of a pet at the hands of another human

laws making "convenience euthanasia" an animal cruelty violation

greater regulation of the pet food industry, including more rigorous inspection of ingredients, greater transparency about sourcing and ingredients, and a well-coordinated method of alerting consumers about recalls

Some of these are more politically and culturally feasible than others, but all are well within the realm of possibility. And each one could help change the climate of vulnerability within which companion animals exist.[2]

47. Speaking for Spot

I said at the beginning of this book that I would use the established vocabulary of pet culture—and particularly the words "pet" and "owner"—until the end because this language most accurately captures where we are with animals at the moment. It would seem just plain odd to speak of "companion animal breeding facilities" or "warehousing of live companions." And how about, "Friend for sale, $200"?

Philosopher Hilary Bok argues that until our behavior changes we need to continue using language that accurately reflects the facts. And the fact is, a good many people who keep animals as pets do not act as guardian or caregiver, in any meaningful respect. They treat their animals like objects or worse. Within the pet industry, animals are bought and sold as commodities. Bok says, "Protesting animals' legal status as property with this terminological change [from 'owner' to 'guardian'] is, in my opinion, as misguided as protesting the existence of slavery by replacing the term 'slave owner' with 'slave caregiver': it obscures the issues it is meant to ameliorate."[1]

Although I agree with Bok, I also think that using different language can help us transform the facts. As Walter Lippmann suggests, we define first and then see.[2] Bit by bit, we transform our seeing and our speaking, and both sight and expression transform under the illumination of empathy. We experience more and more moments of disruption, where the inadequacy of our language grates against our sensibility of who animals are, and we become open to new possibilities.

Here are some ways that activists and philosophers have been trying to change our language, for the benefit of our animal friends. I am offering nothing new here—all of these have been suggested by other people, in other contexts, over the past several decades. But a real shift in our way of talking hasn't really taken hold, so we need to keep chipping away. The linguistic changes are not confined to pets because we need to change how we talk about animals in general. As is often the case when we experiment with new ways of speaking, the words sometimes sound off or unnatural.

Pet. "Pet" is a euphemism. We may happily assume that *pets* are animals who are loved; but they are also animals who are exploited for human economic gain, sexual stimulation, or emotional self-fulfillment. "Pet" functions like "meat" to obscure the animal herself and to make invisible her individuality.

"Pet" is, of course, a term of endearment and maybe it is fine to whisper "my pet" into the ears to our furry friends. But it may be better, in more public settings and in written material, to use an alternative such as "companion animal." Still, even "companion animal" is problematic because it suggests that animals can be classified according to their human uses, so that being a "companion" or being "food" becomes the animal's defining characteristic. For this reason, some people, myself included, like the phrase "animal companion."

I like "companion," and I like "friend," because both of these capture something crucial about the relationship. And as much as I have harped on the overuse of the "pets as family" idea, I like the language of "family." Family is a much healthier metaphor for human-animal relations than object-owner. It is the overuse and cheapening of the pet-as-family idea is that problematic, and not the language itself.

Owner. "Owner" is still technically accurate, but many find it morally offensive and believe that taking ownership over another living being is an act of violence. A handful of U.S. municipalities have changed the term for pet owner to "guardian" or "owner/guardian," a useful shift in legal terminology that could be widely implemented. "Parent" seems to work for many people, but certainly won't appeal to everyone. Companion. Protector. Caregiver (not care*taker*). All of these work, in different contexts. Perhaps the "owner" should choose whichever alternative feels most comfortable. Or, we could stick with "owner," but use obnoxiously obvious air quotes every time we say it, to indicate our dissatisfaction with the concept. At the same time, the fact that Maya is my property isn't all bad. In practical terms, it is a good thing, because nobody can take her from me. As long as animals are considered property they will not gain legal protection as "persons," but their status as property offers a tiny thread of protection.

It. Animals are not genderless objects. "He" and "she" are vast improvements over "it." "Who" is more fitting than "that" or "which." (So, it isn't *what* you are buying at the pet store, but *who* you are buying.)

Bitch and other animal insults. It would be nice if animals weren't used to insult humans. This might seem nitpicking, but think about how racial or gender insults that used to roll off people's tongues sound so jarring now: kike, nigger, cunt. It is hard even to write these in print. Likewise, speciesist insults reinforce stereotypes about animals and humans ("bitch" is both speciesist and sexist) and express our distaste and disdain. It would be nice to see a day when animal insults make people uncomfortable. And if we really feel the need to insult someone, there are plenty of nonanimal alternatives. Other language we might consider for the metaphorical chopping block: the generic "he acted like an animal" (without morals) or "they live like animals" (in filth and squalor) or "she died like an animal" (alone, in pain, with no one to notice her passing).

Euthanasia. This term, if we continue to use it at all, should be reserved for the instances in which we hasten the death of a companion animal to relieve intractable suffering near the end of life. In all other situations, "euthanasia" functions as a euphemism that perpetuates our cultural denial of animal suffering. “Killing” makes people uncomfortable, as it should.

The final piece of verbal activism is this: we need to speak out in defense of companion animals. We are accountable for what we say and also for what we do not say. Silence is a form of acquiescence. *Qui tacet consentire videtur*. He who is silent is taken to agree.

48. So, *Is* Pet Keeping Ethical?

Ethical issues are often presented as a choice between right and wrong: either you are for something or against it. But some ethical quandaries don't fit into this tidy mold, and pet keeping is one of them. I hope that after reading this book you feel less certain about whether pet keeping is ethical than you did on page one. If so, then I've been successful.

Many things about pet keeping should give us pause. Millions of companion animals are suffering, and not just dogs and cats but a diverse range of creatures great and small. By not objecting to what is happening to pet animals—the widespread abuse, the sexual exploitation, the killing of millions of unwanted pets each year, the disposable animal culture, the cruel breeding practices, the high mortality rates of animals being warehoused and sold as pets, the garbage that is marketed as pet food, the boredom and frustration felt by the animals who spend their entire life in solitary confinement—we are giving our tacit assent. By remaining silent, we agree to or choose to ignore the status quo.

Our love for our companion animals can impel us to take a deeper look at just what pet keeping means *to them*. Once we've looked, we need to speak out. And since humans tend to be very hard of hearing when it comes to animal welfare, we need to speak with a clear and loud voice.

We, alongside millions of animals, have been caught up in the tidal wave of pet keeping. Hopefully we've reached the crest of the wave and can work now to slow down the pace of industry growth, reduce the flow of animals, and begin a realistic assessment of the damage. I have no illusions that pet keeping is going away. But my hope is that it will decline in popularity, become far less profitable, and evolve into a more compassionate and animal-friendly affair.

We know how to do it well. The positive aspects of pet keeping are all around us, in the unusual and close friendships that people and animals form and the lengths that people will go to provide a safe and loving home for their animals. My own best argument for pet keeping is right behind me in my office. Bella is curled up on the green-plaid dog bed, and Maya is curled up on the blue dog bed with her head tucked under her leg, like a penguin. Thor is sitting on an office chair carefully grooming between his toes. We need animals and we benefit from our interactions with them, and I would like to think that they—in the best of situations—can benefit from their relations with us. Is it possible to imagine a world in which we can live in close friendship with animals, without causing them harm? I should hope so. Of course, this world is only possible if people have the chance to interact with animals and develop feelings of empathy toward them.

Consider, as an example of empathy building, the way in which Elisabeth Tova Bailey in *The Sound of a Wild Snail Eating* relates to a tiny woodland snail who lives on a plant in her bedroom. She doesn't force the snail into her home—the snail hitches a ride, quite accidentally, on a plant that a friend digs up in the woods and brings to Bailey when she is sick. Through her observations of her molluscan visitor, who remains "free" in his miniature habitat, and by delving into a study of the creature's natural history and behaviors, Bailey awakens to the snail's beauty and to the fascinating world the snail inhabits. The snail remains the master of his own life, and Bailey's response to her pet is so much richer because of it. In time, she returns the snail to the forest, having formed, in her own mind, a close connection and having developed a sense of empathy not only for her "own" snail but for all mollusks and, more broadly, for a previously unnoticed world that exists in the forest outside her home. How much more interesting (and happy) a creature is a snail in his own world than a snail in an artificial environment of our devising. And how much better a teacher of wonder and fellow feeling.

The human desire to associate with other animals, to observe them and get inside their worlds, lies at the heart of pet keeping. And this impulse to connect seems vitally important to nurture in ourselves and in our children. Maybe what we need to challenge is not pet keeping, per se, but some of the forms that pet keeping takes. The model of pet keeping most familiar to us, and that promoted by the pet industry—an animal in a cage, who we take out and observe or stroke and then put back—is only one possibility. And in fact, maybe it is a form of pet keeping that fails to satisfy anybody's interests, with the exception of those making money on the sale of pets and pet products. We are likely to find caged animals uninteresting—all they do (all they can do) is just sit there. It is no wonder that children become bored with their hamsters and goldfish. If we were to find natural points of overlap between animals' worlds and ours, we might enjoy encounters that are less artificial and more dynamic because they preserve the integrity of the animals themselves.

I leave you with a call to action. Change starts with awareness. By increasing awareness about the issues raised in this book, ranging from personal pet ownership to the global pet industry, we can begin to make more careful decisions about our own pets and about the broader community of animals with whom we live. Through education, careful decision making about animal guardianship, and active efforts to thwart abuse, neglect, and abandonment, we can all make a difference. Awareness also gives birth to compassion, which may call some of us on to advocacy on a broader scale. Both action and inaction surrounding our pets and the pet industry will have consequences. The question I leave you with is: How do you wish to define your role in all this and what changes do you hope to impart?

Notes

Full publication information for references cited in the notes below can be found in the bibliography that follows.

CHAPTER 3

1. For an overview of domestication, see Francis, *Domesticated*.
2. Daniels and Bekoff, "Domestication, Exploitation, and Rights."
3. On dogs, see, e.g., Derr, *How the Dog Became the Dog*; and Range and Virányi, "Tracking the Evolutionary Origins of Dog-Human Cooperation: The 'Canine Cooperation Hypothesis.'" On cats, see Hua et al., "Earliest Evidence for Commensal Processes of Cat Domestication"; and Driscoll et al. "The Taming of the Cat."
4. Wilkins, Wrangham, and Fitch, "The 'Domestication Syndrome."

CHAPTER 4

1. See, e.g., DeLoache, Pickard, and LoBue, "How Very Young Children Think"; and Melson, *Where the Wild Things Are*.
2. Serpell, *In the Company of Animals*, and "Pet-Keeping and Animal Domestication." Serpell's claim may be overstated. See, e.g., Herzog, "Biology, Culture, and the Origins of Pet-Keeping."
3. Serpell, "Pet-Keeping and Animal Domestication," 13.
4. Ibid.
5. Julius et al., *Attachment to Pets*, 32–33.
6. Tuan, *Dominance and Affection*, 1–2.
7. Ibid., 5.

CHAPTER 6

1. Barker and Barker, "Human-Canine Bond."
2. Olin-Vertesh, "Flexible Personhood."
3. Nagasawa et al., "Oxytocin-Gaze Positive Loop and the Coevolution of Human-Dog Bonds"; Archer and Monton, "Preferences for Infant Facial Features in

Pet Dogs and Cats"; and Bradshaw and Paul, "Could Empathy for Animals Have Been an Adaptation in the Evolution of *Homo sapiens*?"

4. Beck and Katcher, *Between Pets and People*, 57.
5. Robin and ten Bensel, "Pets and the Socialization of Children."
6. "Woman Misses the Affection That Fiancé Shows to Pets," Dear Abby, *Longmont Times-Call*, March 2, 2015.
7. See, e.g., Bowen, *Family Therapy*.
8. Harker, Collis, and McNicholas, "Influence of Current Relationships."
9. According to a recent review of the empirical data, there is little support for the hypothesis that companion animals help alleviate loneliness. Most of the studies claiming that animals help treat loneliness have been flawed or underpowered. Gilbey and Tani, "Companion Animals and Loneliness."

CHAPTER 7

1. Mellor, Patterson-Kane, and Stafford, *Sciences of Animal Welfare*, 125.
2. Alderman, *The Book of Times*, 18.

CHAPTER 8

1. Quoted in de Laroche and Labat, *Secret Life of Cats*, 86.
2. Mayo Clinic, "Are Your Pets Disturbing Your Sleep? You're Not Alone," *ScienceDaily*, June 3, 2014, www.sciencedaily.com/releases/2014/06/140603193830.htm.
3. The fact that some dogs behave aggressively in bed is incontrovertible. The explanation for bed-related aggressive behavior, however, is controversial. Many behaviorists assert that dogs allowed on the bed get confused about pack hierarchy and may get an inflated idea about their status in the household. This points to a very interesting discussion about whether "pack hierarchy" is an appropriate metaphor for human-canid social relationships, and how the term "dominance" should be understood. See, e.g., the American Society for the Prevention of Cruelty to Animals' discussion of dominance, hierarchy, and aggression: "Is Your Dog Dominant?" https://www.aspca.org/pet-care/virtual-pet-behaviorist/dog-behavior/your-dog-dominant; and on "dominance" as explanation for behavior, see also American Veterinary Society of Animal Behavior's "Position Statement on the Use of Dominance Theory in Behavior Modification of Animals," http://avsabonline.org/uploads/position_statements/Dominance_Position_Statement_download-10-3-14.pdf, 2008; and Welfare in Dog Training's "What's Wrong with Using 'Dominance' to Explain the Behavior of Dogs," http://www.dogwelfarecampaign.org/why-not-dominance.php

CHAPTER 9

1. Vrontou et al., "Genetic Identification of C Fibres."
2. Ramos et al., "Are Cats More Stressed?"

CHAPTER 10

1. Meints, Racca, and Hickey, "How to Prevent Dog Bit Injuries."
2. See, e.g., Hare and Tomasello, "Human-Like Social Skills"; and Téglás et al., "Dog's Gaze Following."
3. Takaoka et al., "Do Dogs Follow Behavioral Cues from an Unreliable Human?" The bibliographic references in this article are worth perusing for fascinating information about dog-human communication and dog social cognition.
4. Pongrácz et al., "Do Children Understand."
5. Mariti et al., "Perception of Dogs' Stress."
6. See, e.g., Mills, van der Zee, and Zulch, "When the Bond Goes Wrong," 235.
7. Ibid., 236.
8. Grandin and Johnson, *Animals Make Us Human*, 74.
9. For a nice summary of rat behavior, see Anne Hanson, "Glossary of Rat Behavior Terms," last updated June 10, 2008, http://www.ratbehavior.org/Glossary.htm.
10. Williams, *Ask Your Animal*, 35.

CHAPTER 11

1. American Pet Products Association, "Pet Industry Market Size and Ownership Statistics," http://www.americanpetproducts.org/press_industrytrends.asp.

CHAPTER 13

1. DeLoache, Pickard, and LoBue, "How Very Young Children Think"; and Melson, *Why the Wild Things Are*.
2. Lorenz, "Waning of Humaneness," 208, quoted in Serpell and Paul, "Pets and the Development of Positive Attitude," 138.
3. For a nice but somewhat dated overview of the research on pets and child development, see Endenburg and Baarda, "The Role of Pet in Enhancing Human Well-Being: Effects on Child Development."
4. Mueller, "Is Human-Animal Interaction?"; see also *ScienceDaily*'s article about the study: Tufts University, "Caring for Animals May Correlate with Positive Traits in Young Adults," *ScienceDaily*, January 31, 2014, http://www.sciencedaily.com/releases/2014/01/140131230731.htm.
5. Poresky, "Young Children's Empathy"; see also Beetz, "Empathy as an Indicator," 50.

6. Serpell and Paul, "Pets and the Development of Positive Attitude," 139.
7. See, e.g., Paul, "Empathy with Animals and with Humans: Are They Linked?"

CHAPTER 14

1. Alexandra Sifferlin, "Why Having a Dog Helps Keep Kids Asthma-Free," *Time*, June 20, 2012, http://healthland.time.com/2012/06/20/why-having-a-dog-may-keep-kids-asthma-free; Cesar Millan, "Your Dog Can Be the Secret to Weight Loss," Cesar's Way, http://www.cesarsway.com/dog-training/exercise/Your-Dog-Can-Be-The-Secret-To-Weight-Loss; Jeanie Lerche Davis, "Five Ways Pets Can Improve Your Health," WebMD Feature Archive, http://www.webmd.com/hypertension-high-blood-pressure/features/health-benefits-of-pets)
2. Headley and Grabka, "Health Correlates of Pet Ownership."
3. Ibid., 155.
4. Selhub and Logan, *Your Brain on Nature*, 146.

CHAPTER 15

1. For a recent review, see Stull, Jason, Brophy, and Weese, "Reducing the Risk of Pet-Associated Zoonotic Infections."
2. Hanauer, Ramakrishnan, and Seyfried, "Describing the Relationships between Cat Bites and Human Depression."
3. Weese and Fulford, *Companion Animal Zoonoses*, viii.
4. Here are some resources, if you want to learn more: Companion Animal Parasite Council ("CAPC Recommendations," http://www.capcvet.org/capc-recommendations), National Collaborating Centre for Environmental Health (Angela Smith and Yvonne Whitfield, "Household Pets and Zoonoses," January 2012, http://ncceh.ca/sites/default/files/Household_Pets_Zoonoses_Jan_2012.pdf), and Centers for Disease Control ("Keeping Pets Healthy Keeps People Healthy Too!" last updated July 1, 2015, http://www.cdc.gov/healthypets/).

CHAPTER 16

1. Some legal scholars would like to see denial of necessary veterinary care an animal cruelty violation. See, e.g., Hankin, "Making Decisions about Our Animals' Health Care"; and Coleman, "Man['s Best Friend] Does Not Live by Bread Alone."

CHAPTER 17

1. Bradshaw, "The Evolutionary Basis for Feeding Behavior."
2. Thurston, *Lost History*.
3. Tegzes et al., *Just Food for Dogs*, 8.
4. Michael Myers, "CVM Scientists Develop PCR Test to Determine Source of

Animal Products in Feed, Pet Food," *FDA Veterinarian Newsletter* 19, no. 1 (January/February 2004), https://archive.is/uEie.

5. Kawalek et al., "Effect of Oral Administration."

CHAPTER 19

1. McMillan, "Stress-Induced Emotional Eating."

CHAPTER 20

1. Cinquepalmi et al., "Environmental Contamination by Dog Faeces," 72.
2. Lowe et al., "Environmental and Social Impacts of Dog Waste."

CHAPTER 21

1. Centers for Disease Control and Prevention, "Preventing Dog Bites," last updated May 18, 2015, http://www.cdc.gov/features/dog-bite-prevention/index.html; American Veterinary Medical Association, "Infographic: Dog Bites by the Numbers," 2015, https://www.avma.org/Events/pethealth/Pages/Infographic-Dog-Bites-Numbers.aspx.
2. To get a sense of the spit and fire surrounding the issue of pit bulls and dangerousness, you can simply Google "pit bull." For specific examples, compare the research presented on the Animal Farm Foundation website (http://www.animalfarmfoundation.org/pages/Breed-Specific-Legislation) with Merritt Clifton's "Dog Attack Deaths and Maimings, U.S. and Canada September 1982 to November 13, 2006" (http://dogbitelaw.com/images/pdf/Dog_Attacks_1982-2006_Clifton.pdf). On the unreliability of visual breed identification by shelter personnel, research by Dr. Victoria Voith is particularly pertinent: http://nationalcanineresearchcouncil.com/breed-identification-1/.
3. Lakestani, Donalson, and Waran, "Interpretation of Dog Behavior."
4. Meints, Racca, and Hickey, "How to Prevent Dog Bit Injuries."
5. Robin Bennett's website has excellent information on keeping kids and dogs safe: "Why Supervising Dogs and Kids Doesn't Work," August 19, 2013, http://www.robinkbennett.com/2013/08/19/why-supervising-dogs-and-kids-doesnt-work/.

CHAPTER 22

1. Bukowski and Wartenberg, "An Alternative Approach."
2. "Breaking New Ground on Toxins in Pets," Environmental Working Group, http://www.ewg.org/successes/2008/breaking-new-ground-toxins-pets.
3. Norrgran et al., "Higher PBDE Serum Concentrations."
4. De Silva and Turchini, "Towards Understanding the Impacts."
5. Estimates of number of pets: "Pet Industry Market Size and Ownership Statistics," http://www.americanpetproducts.org/press_industrytrends.asp;

cell phone numbers are from an industry survey, reported in the *Washington Post* (Cecilia Kang, "Number of Cellphones Exceeds U.S. Population: CTIA Trade Group," *Technology* [blog], *Washington Post*, October 11, 2011, http://www.washingtonpost.com/blogs/post-tech/post/number-of-cell-phones-exceeds-us-population-ctia-trade-group/2011/10/11/gIQARNcEcL_blog.html).
6. Ghirlanda, Acerbi, and Herzog, "A Case Study in Media Influence on Choice."
7. Bush, Baker, and MacDonald, "Global Trade in Exotics."
8. Baker et al., "Rough Trade."

CHAPTER 23

1. These data are from the American Pet Products Association. Numbers are also available for various exotic species. The American Veterinary Medical Association gives lower estimates: 70 million dogs; 75 million cats; 8 million birds, etc. The AVMA data is from 2012, while APPA is from 2014. But the time lag is probably not enough to account for the discrepancy between the numbers. Without detailed information about how the data were collected by each organization, it is hard to say which is more accurate and why they are so different. Could APPA figures be bloated, to oversell the popularity of pets as products? APPA data can be found here: "Pet Industry Market Size and Ownership Statistics," http://www.americanpetproducts.org/press_industrytrends.asp. AVMA numbers are here: "U.S. Pet Ownership Statistics," https://www.avma.org/KB/Resources/Statistics/Pages/Market-research-statistics-US-pet-ownership.aspx.
2. Warwick, "The Morality of the Reptile Trade," 79.
3. Ibid., for a list of additional references.
4. Kis, Huber, and Wilkinson, "Social Learning by Imitation."
5. Warwick, "The Morality of the Reptile Trade," 78.
6. Ibid., 79.
7. Mazorlig, "Small Tanks," 92–93.
8. Aydinonat et al., "Social Isolation."
9. Douglas Quenqua, "An Idyllic Picture of Serenity, But Only If You're Not Inside," *New York Times*, December 26, 2011, http://www.nytimes.com/2011/12/27/science/fish-in-small-tanks-are-shown-to-be-much-more-aggressive.html.
10. Horowitz, "*Canis familiaris*."
11. Ibid., 11–12.
12. Ibid., 13.
13. Ibid., 16.
14. See Palmer and Sandøe, "For Their Own Good."
15. Sonntag and Overall, "Key Determinants," 217.

CHAPTER 24

1. Wemelsfelder, "Animal Boredom," 81.
2. Ibid., 83.
3. Ibid., 84.
4. Ibid., 81.
5. Ibid.
6. Ibid., 85.

CHAPTER 25

1. Jonathan Bullington, "Niles Man Left Dog Unattended for Almost Two Months," *Chicago Tribune*, March 19, 2014, http://articles.chicagotribune.com/2014-03-19/news/ct-niles-animal-cruelty-arrest-tl-20140319_1_niles-man-niles-police-news-release-apartment-building.
2. The link between attachment and separation is complicated, and although there is no question that many of the anxiety problems suffered by our nations' population of dogs are rooted in separation anxiety, caution is in order when interpreting the root cause of a dog's behavior. As Mills, van der Zee, and Zulch suggest, "separation anxiety" makes an inference about the cognitive and affective processes underlying the problem. But the inference may not be correct. For example, a dog who systematically chews every piece of furniture in the house when his owners are gone may have separation anxiety. He may, alternatively, be motivated by frustration associated with being confined (by what Jaak Panksepp would call the RAGE sensu, in *Affective Neuroscience*). Destruction may not be associated with distress at all but, rather, with the positive SEEKING affective system. Mills, van der Zee, and Zulch, "When the Bond Goes Wrong," 235.
3. Previde and Valsecchi, "The Immaterial Cord," 178.
4. Ibid.

CHAPTER 26

1. Emily Thomas, "This Collie Lost Lower Jaw, Nearly Died from Gunshot Wound." *Huffington Post*, March 21, 2014, http://www.huffingtonpost.com/2014/03/21/collie-lower-jaw-lad-gunshot-_n_5008200.html.
2. "Man Tries to Get Dog to Bite Bystanders by Kicking and Punching Dog," Examiner.com, February 12, 2014, http://www.examiner.com/article/man-tries-to-get-dog-to-bite-bystanders-by-kicking-and-punching-dog?CID=examiner_alerts_article.
3. "NJ Man Charged after Dragging Dog Tied to Car," *NBC New York*, March 31, 2014, http://www.nbcnewyork.com/news/local/Man-Charged-After-Dragging-Dog-Tied-to-Car-New-Jersey-253102191.html.
4. "Man Says He Killed Dog Because She 'Looked at Him Funny,'" *First Coast*

News, March 7, 2014, http://www.firstcoastnews.com/story/news/local/2014/03/07/man-kills-adopted-dog/6172801/.

5. Sinclair, Merck, and Lockwood, *Forensic Investigation*, 1.
6. Revised Code of Washington, sec. 16.52.205, "Animal Cruelty in the First Degree," http://apps.leg.wa.gov/rcw/default.aspx?cite=16.52.205.
7. Jessica Wilder, "Man Accused of Blowing up Dog Not Charged with Animal Cruelty," ABC News, August 6, 2013, http://abcnews.go.com/blogs/headlines/2013/08/man-accused-of-blowing-up-dog-not-charged-with-animal-cruelty/.
8. Sinclair, Merck, and Lockwood, *Forensic Investigation*, 21.
9. Vermeulen and Odendaal, "Proposed Typology," 7.
10. Carlisle-Frank and Flanagan, *Silent Victims*, 79.
11. Ibid., 77.
12. Ibid., 7.
13. Mark Derr, "When Did You Stop Kicking, Hitting Your Spouse, Dog, Child?" *Psychology Today* (blog), October 22, 2014, https://www.psychologytoday.com/blog/dogs-best-friend/201410/when-did-you-stop-kicking-hitting-your-spouse-dog-child.

CHAPTER 27

1. Frank McMillan has been one of the most vocal advocates for animals in regard to emotional suffering and well-being. See, e.g., McMillan, "Emotional Maltreatment," 174.
2. Ibid., 171–72.

CHAPTER 29

1. Patronek, "Animal Hoarding," 222.
2. Ibid., 227.
3. Ibid., 221.

CHAPTER 30

1. Akhtar, *Animals and Public Health*.
2. See also Gullone, "An Evaluative Review," for a recent review of the research.
3. Akhtar, *Animals and Public Health*, 30.
4. Phil Arkow, "The 'Link' with Domestic Violence," http://animaltherapy.net/animal-abuse-human-violence/the-link-with-domestic-violence/.
5. Akhtar, *Animals and Public Health*, 39–42.

CHAPTER 31

1. Jenny Edwards, "Webinar: Understanding and Prosecuting Bestiality," ASPCA Professional, January 14, 2014, http://www.aspcapro.org/webinar/2014-01-14/understanding-and-prosecuting-bestiality.

2. Beetz, "Beastiality and Zoophilia: A Discussion," 204; Dekkers, *Dearest Pet.*
3. Ibid., 216.
4. Bolliger and Goetschel, "Sexual Relations with Animals," 25.
5. For example, Hawthorne, *Bleating Hearts.*
6. Edwards, "Webinar: Understanding and Prosecuting Bestiality," http://www.aspcapro.org/webinar/2014-01-14/understanding-and-prosecuting-bestiality.
7. Munro and Thrusfield, "Battered Pets," 335.
8. Merck and Miller, "Sexual Abuse," 233.
9. Beetz, "Bestiality and Zoophilia: Associations," 204.
10. Beck and Katcher, *Between Pets and People*, 56.
11. Ascione, *Children and Animals*, 126.

CHAPTER 32

1. Lesy, *The Forbidden Zone*, 4.
2. A summary of state laws governing euthanasia can be found here: "Summary of Each State's Laws on Euthanasia," American Veterinary Medical Foundation, last updated May 5, 2015, https://www.avma.org/Advocacy/StateAndLocal/Documents/euthanasia_laws.pdf.

CHAPTER 33

1. "Destroying the Dogs," *New York Times*, July 6, 1877.
2. Larry Carbone forcefully makes this argument (*What Animals Want*, 202).
3. Various philosophers have written about the better-off-dead question and the harm of death. These discussions are interesting and useful. See, e.g., Carruthers, *The Animals Issue*; Regan, *The Case for Animal Rights*; DeGrazia, *Taking Animals Seriously*; and Sapontzis, *Morals, Reasons, and Animals.*
4. Wright, "Why Must We Euthanize," 7–8.
5. McMahan, *The Ethics of Killing*, esp. 199–203.
6. Ibid., 203.
7. Brestrup, *Disposable Animals*, 44.

CHAPTER 34

1. Pierce, *The Last Walk.*

CHAPTER 35

1. "Pets by the Numbers," Humane Society of the United States, January 30, 2014, http://www.humanesociety.org/issues/pet_overpopulation/facts/pet_ownership_statistics.html.
2. Hoffman, Creevy, and Promislow, "Reproductive Capability."
3. Kustritz, "Determining the Optimal Age."
4. Torres de la Riva et al., "Neutering Dogs."

5. Milani, "Canine Surgical Sterilization," 924.
6. Ibid.
7. These numbers are drawn, by Peter Marsh, from Merritt Clifton's work: "Replacing Myth with Math: Using Data to Design Shelter Overpopulation Programs," chap. 1, 2010, http://www.shelteroverpopulation.org/SOS_Chapter-1.pdf.
8. Kustritz, "Determining the Optimal Age."
9. Rollin, *Animal Rights*.
10. See Peeters and Kirpensteijn, "Comparison of Surgical Variables"; and DeTora and McCarthy, "Ovariohysterectomy versus Ovariectomy."
11. Alliance for Contraception in Cats and Dogs, "Zeuterin™/Esterisol™: Product Profile and Position Paper," June 2014," http://www.acc-d.org/docs/default-source/Research-and-Innovation/pppp-zeuterinesterilsol-revised-june2015-for-web.pdf?sfvrsn=2.
12. Armbruster, "Into the Wild," 43, 46.

CHAPTER 36

1. "Puppy Mill FAQ," American Society for the Prevention of Cruelty to Animals, http://www.aspca.org/fight-cruelty/puppy-mills/puppy-mill-faq.
2. "What Everyone Needs to Know about Hunte," http://www.thehuntecorporation.com/.
3. "Undercover at the Hunte Corporation," Companion Animal Protection Society, 2004, http://www.caps-web.org/rescues/item/656-undercover-at-the-hunte-corporation.
4. Brandow, *A Matter of Breeding*.

CHAPTER 37

1. Statistics on shelter killings are notoriously difficult to interpret. In 2014, the latest year for which data are available, the number of dogs and cats killed appears to have dropped to 2.7 million, the lowest on record in sixty years. But the lower numbers may simply represent a shifting around of where animals are dying and who is doing the killing. For example, feral cats disposed of by the growing army of "wildlife nuisance" companies, in lieu of being trapped and taken to a shelter to be killed, generally go unreported. Merritt Clifton's Animals-24/7 website is a good resource on shelter statistics. For these particular statistics, see Merritt Clifton, "Record Low Shelter Killing Raises Both Hopes and Questions," November 14, 2014, http://www.animals24–7.0rg/2014/11/14/record-low-shelter-killing-raises-both-hopes-questions/.
2. Paul Leach, "Dogs 'Rescued' from Cleveland Shelter Sent to 'House of Horrors,'" *Chattanooga Times Free Press*, February 20, 2014, http://www.timesfree

press.com/news/local/story/2014/feb/20/dogs-from-cleveland-rescue-seized-in-morristown/132243/.

3. "Executive Director of ARF Found Dead in Car with 31 Dogs," Examiner.com, November 8, 2013, http://www.examiner.com/article/executive-director-of-arf-found-dead-car-with-31-dogs?cid=taboola_inbound.
4. Leigh and Geyer, *One at a Time*, 52.
5. Ibid., 53.
6. Ibid., 52.
7. Ibid.
8. "Pet Care Costs," American Society for the Prevention of Cruelty to Animals," https://www.aspca.org/adopt/pet-care-costs.
9. "Man Found Traveling with 62 Dogs by Minivan," January 2014, http://www.nbc4i.com/story/24429277/man-found-traveling-with-62-dogs-by-minivan (page no longer available).

CHAPTER 38

1. "To Hell and Back: A Journey Inside the Pet Trade," PETA, http://features.peta.org/pettrade/.
2. Ashley et al., "Mortality and Morbidity," 308.
3. Ibid., 317.

CHAPTER 39

1. Boncy, "What Tomorrow Brings," 54.
2. Ibid., 60.
3. Ibid., 59–60.
4. Ibid., 54.
5. Ibid., 59.
6. Ibid., 61. To this end, various pet industry groups such as PIJAC and American Pet Products Association, as well as large companies like PetCo and Zoetis (which sells veterinary pharmaceuticals), are offering financial support to the Human Animal Bond Research Initiative, which studies and promotes the purported health benefits of pets.

CHAPTER 40

1. "Your Freedom to Responsibly Own Fish Is at Risk," PIJAC, updated March 25, 2015, http://www.pijac.org/marine.
2. "Dog Ownership . . . a Cherished American Tradition," National Animal Interest Alliance, October 3, 2013, http://www.naiaonline.org/articles/article/dog-ownership . . .-a-cherished-american-tradition.
3. "Frequently Asked Questions," Protect the Harvest, http://protecttheharvest.com/who-we-are/faqs/.

CHAPTER 41

1. Hannah: The Pet Society, http://www.hannahsociety.com/.
2. William Grimes, "Cat Café Offers a Place to Snuggle, with Reservations," *New York Times*, January 16, 2015, http://www.nytimes.com/2015/01/16/nyregion/cat-cafe-offers-a-place-to-snuggle-with-reservations.html?_r=0.

CHAPTER 42

1. Bush, Baker, and Macdonald, "Global Trade in Exotics," 665.
2. Ibid.
3. Ibid.

CHAPTER 43

1. Špinka and Wemelsfelder, "Environmental Challenge."
2. Ibid., 28.
3. Ibid., 27.

CHAPTER 44

1. Young, *Environmental Enrichment*, 76.
2. Ibid.
3. Ibid., 77.
4. This definition comes from Wemelsfelder, "Animal Boredom," 87.
5. Ibid., 87.
6. Young, *Environmental Enrichment*, 79.
7. Braithwaite, *Do Fish Feel Pain?*
8. Salvanes et al., "Environmental Enrichment Promotes."
9. Karen Pryor's Clicker Training website: Karen Pryor, "Fish Enrichment," Karen Pryor Clicker Training, December 1, 2004, http://www.clickertraining.com/node/29.
10. Barbara Heidenreich, "Cute Guinea Pigs Play Basketball," December 22, 2013, https://www.youtube.com/watch?v=0461ujOgBy8#t=81.

CHAPTER 45

1. Burghardt, "Environmental Enrichment."
2. FooPets: http://www.foopets.com/.

CHAPTER 46

1. Faunalytics, "Animal Tracker, Year 6," 2013, https://faunalytics.org/feature-article/animal-tracker-year-6/#. (The full report is available to registered users only).
2. On legal changes, Phillips, *Defending the Defenseless*; Animal Legal Defense

Fund (http://aldf.org/); Born Free USA (http://www.bornfreeusa.org/); and Companion Animal Protection Society http://www.caps-web.org/).

CHAPTER 47

1. Bok, “Keeping Pets,” 791.
2. Lippmann, *Public Opinion*.

Bibliography

Aaltola, Elisa. "Animal Suffering: Representations and the Act of Looking." *Anthrozoös* 27 (2014): 19–31.

———. *Animal Suffering: Philosophy and Culture*. New York: Palgrave Macmillan, 2012.

Adams, Carol. *The Sexual Politics of Meat: A Feminist-Vegetarian Critical Theory*. Rev. ed. London: Bloomsbury Academic, 2010.

Akhtar, Aysha. *Animals and Public Health: Why Treating Animals Better Is Critical to Human Welfare*. New York: Palgrave Macmillan, 2012.

Alderman, Lesley. *The Book of Times*. New York: William Morrow, 2013.

Alderton, David. *A Petkeeper's Guide to Hamsters and Gerbils*. London: Salamander Books, 1996.

Alger, Janet M., and Steven F. Alger. *Cat Culture: The Social World of a Cat Shelter*. Philadelphia: Temple University Press, 2003.

American College of Veterinary Behaviorists. *Decoding Your Dog: The Ultimate Experts Explain Common Dog Behaviors and Reveal How to Prevent or Change Unwanted Ones*, edited by Debra F. Horwitz, John Ciribassi, with Steve Dale. Boston: Houghton Mifflin Harcourt, 2014.

American Humane Association. *Euthanasia by Injection: Training Guide*. Washington, DC: American Humane Association, 2011.

———. *Operational Guide for Animal Care and Control Agencies: Euthanasia by Injection*. Washington, DC: American Humane Association, 2010.

American Veterinary Medical Association, Task Force on Canine Aggression and Human-Canine Interactions. "A Community Approach to Dog Bite Prevention." *Journal of the American Veterinary Medical Association* 218 (2001): 1732–49.

American Veterinary Medical Association. *AVMA Guidelines for the Euthanasia of Animals: 2013 Edition*. Schaumburg, IL: AVMA, 2013.

American Veterinary Society of Animal Behavior. "Position Statement on the Use of Dominance Theory in Behavior Modification of Animals." 2008. http://

avsabonline.org/uploads/position_statements/Dominance_Position_Statement_download-10-3-14.pdf.

Amiot, C., and B. Bastain. "Toward a Psychology of Human-Animal Relations." *Psychological Bulletin* 141, no. 1 (2014): 1–42.

Anderson, David C. *Assessing the Human-Animal Bond: A Compendium of Actual Measures*. West Lafayette, IN: Purdue University Press, 2007.

Anderson, Elizabeth P. *The Powerful Bond between People and Pets: Our Boundless Connections to Companion Animals*. Westport, CT: Praeger, 2008.

Anderson, Patricia K. "Social Dimensions of the Human-Avian Bond: Parrots and Their Persons." *Anthrozoös* 27 (2014): 371–87.

Animal Welfare Institute. *The Animal Dealers: Evidence of Abuse of Animals in the Commercial Trade, 1952–1997*. Washington, DC: Animal Welfare Institute, 1997.

Appleby, Michael C., Joy A. Mench, I. Anna S. Olsson, and Barry O. Hughes, eds. *Animal Welfare*. 2nd ed. Cambridge, MA: CAB, 2011.

Archer, John. "Pet Keeping: A Case Study in Maladaptive Behavior." In *The Oxford Handbook of Evolutionary Family Psychology*, edited by Catherine A. Salmon and Todd K. Shackelford, 281–96. New York: Oxford University Press, 2011.

———. "Why Do People Love Their Pets?" *Evolution and Human Behavior* 18 (1997): 237–59.

Archer, John, and S. Monton. "Preferences for Infant Facial Features in Pet Dogs and Cats." *Ethology* 117 (2011): 237–59.

Arluke, Arnold. *Just a Dog: Understanding Animal Cruelty and Ourselves*. Philadelphia: Temple University Press, 2006.

Arluke, Arnold, and Clinton R. Sanders. *Regarding Animals*. Philadelphia: Temple University Press, 1996.

Armbruster, Karla. "Into the Wild: An Ecofeminist Perspective on the Human Control of Canine Sexuality and Reproduction." In *Ecofeminism and Rhetoric: Critical Perspectives on Sex, Technology, and Discourse*, edited by Douglas A. Vakoch, 39–64. New York: Berghahn Books, 2011.

Ascione, Frank. "Animal/Pet Abuse" In *Encyclopedia of Interpersonal Violence*, edited by Claire M. Renzetti and Jeffrey Edleson, 27–28. Los Angeles: SAGE Publications, 2008.

———. "Bestiality." In *Encyclopedia of Interpersonal Violence*, edited by Claire M. Renzetti and Jeffrey Edleson, 76–77. Los Angeles: SAGE Publications, 2008.

———. "Bestiality: Petting, 'Humane Rape,' Sexual Assault, and the Enigma of Sexual Interactions between Humans and Non-Human Animals." In *Bestiality and Zoophilia: Sexual Relations with Animals*, edited by Andrea M. Beetz and Anthony L. Podberscek, 120–29. West Lafayette, IN: Purdue University Press, 2005.

———. *Children and Animals: Exploring the Roots of Kindness and Cruelty*. West Lafayette, IN: Purdue University Press, 2005.

Ashley, Shawn, Susan Brown, Joel Ledford, Janet Martin, Ann-Elizabeth Nash, Amanda Terry, Tim Tristan, and Clifford Warwick. "Mortality and Morbidity of Invertebrates, Amphibians, Reptiles, and Mammals at a Major Exotic Animal Wholesaler." *Journal of Applied Animal Welfare Science* 17 (2014): 308–21.

Aydinonat, Denise, Dustin J. Penn, Steve Smith, Yoshan Moodley, Franz Hoelzl, Felix Knauer, and Franz Schwarzenberger. "Social Isolation Shortens Telomeres in African Grey Parrots (*Psittacus erithacus erithacus*)." *PLOS One* (April 4, 2014). http://www.plosone.org/article/info%3Adoi%2F10.1371%2Fjournal.pone.0093839.

Baker, Sandra E., Russ Cain, Freya van Kesteren, Zinta A. Zommers, Neil D'Cruze, and David W. Macdonald. "Rough Trade: Animal Welfare in the Global Wildlife Trade." *Bioscience* 63 (2013):928–38.

Bamberger, Michelle, and Robert Oswald. *The Real Cost of Fracking: How America's Shale Boom Is Threatening Our Families, Pets and Food*. Boston: Beacon Press, 2014.

Barker, Sandra B., and Randolph T. Barker. "The Human-Canine Bond: Closer Than Family Ties." *Journal of Mental Health Counseling* 10 (1988): 46–56.

Barman, C. R., N. S. Barman, M. L. Cox, K. B. Newhouse, and M. J. Goldston. "Students' Ideas about Animals: Results from a National Study." *Science and Children* 38 (1991): 42–46.

Beck, Alan, and Aaron Katcher. *Between Pets and People: The Importance of Animal Companionship*. West Lafayette, IN: Purdue University Press, 1996.

Beetz, Andrea M. "Bestiality and Zoophilia: Associations with Violence and Sex Offending." In *Bestiality and Zoophilia: Sexual Relations with Animals*, edited by Andrea M. Beetz and Anthony L. Podberscek, 46–70. West Lafayette, IN: Purdue University Press, 2005.

———. "Bestiality and Zoophilia: A Discussion of Sexual Contact with Animals." In *The International Handbook of Animal Abuse and Cruelty: Theory, Research, and Application*, edited by Frank R. Ascione, 201–20. West Lafayette, IN: Purdue University Press, 2008.

———. "Empathy as an Indicator of Emotional Development." In *The Link between Animal Abuse and Human Violence*, edited by Andrew Linzey, 62–74. Portland, OR: Sussex Academic Press, 2009.

———. "New Insights into Bestiality and Zoophilia." In *Bestiality and Zoophilia: Sexual Relations with Animals*, edited by Andrea M. Beetz and Anthony L. Podberscek, 98–119. West Lafayette, IN: Purdue University Press, 2005.

Beirne, Piers. *Confronting Animal Abuse: Law, Criminology, and Human-Animal Relationships*. Lanham, MD: Rowman & Littlefield, 2009.

———. "Rethinking Bestiality: Towards a Concept of Interspecies Sexual Assault." In *Companion Animals and Us: Exploring the Relationships between People and Pets*, edited by Anthony L. Podberscek, Elizabeth S. Paul, and James A. Serpell, 313–31. New York: Cambridge University Press, 2000.

Bekoff, Marc. *The Emotional Lives of Animals*. Novato, CA: New World Library, 2008.

———. "The Question of Animal Emotions: An Ethological Perspective." In *Mental Health and Well-Being in Animals*, edited by Franklin D. McMillan, 15–28. Ames, IA: Blackwell Publishing Professional, 2005.

Bekoff, Marc, and Jessica Pierce. *Wild Justice: The Moral Lives of Animals*. Chicago: University of Chicago Press, 2009.

Benning, Lee Edwards. *The Pet Profiteers: The Exploitation of Pet Owners—and Pets—in America*. New York: Quadrangle/The New York Times Book Co., 1976.

Berger, John. *Why Look at Animals?* New York: Penguin Books, 2009.

Berns, Gregory. *How Dogs Love Us: A Neuroscientist and His Adopted Dog Decode the Canine Brain*. New York: New Harvest, 2013.

Bini, John K., Stephen M. Cohn, Shirley M. Acosta, Marilyn J. McFarland, Mark T. Muir, and Joel E. Michalek, for the TRISAT Clinical Trials Group. "Mortality, Mauling, and Maiming by Vicious Dogs." *Annals of Surgery* 253, no. 4 (April 2011): 791–97.

Bok, Hilary. "Keeping Pets." In *The Oxford Handbook of Animal Ethics*, edited by Tom L. Beauchamp and R. G. Frey, 769–95. New York: Oxford University Press, 2011.

Bolliger, Gieri, and Antoine F. Goetschel, "Sexual Relations with Animals (Zoophilia): An Unrecognized Problem in Animal Welfare Legislation." In *Bestiality and Zoophilia: Sexual Relations with Animals*, edited by Andrea M. Beetz and Anthony L. Podberscek, 23–45. West Lafayette, IN: Purdue University Press, 2005.

Bonas, Sheila, June McNicholas, and Glyn M. Collis. "Pets in the Network of Family Relationships: An Empirical Study." In *Companion Animals and Us: Exploring the Relationships between People and Pets*, edited by Anthony L. Podberscek, Elizabeth S. Paul, and James A. Serpell, 209–36. New York: Cambridge University Press, 2000.

Boncy, Jennifer. "What Tomorrow Brings." *Pet Business* February (2014), 54–61.

Born Free USA. "What's really in Pet Food?" 2007. http://www.bornfreeusa.org/facts.php?p=359&more=1.

Bowen, Murray. *Family Therapy in Clinical Practice*. New York: Jacob Aronson, 1978.

Bradshaw, John W. S. *Cat Sense: How the New Feline Science Can Make You a Better Friend to Your Pet*. New York: Basic Books, 2014.

———. "The Evolutionary Basis for the Feeding Behavior of Domestic Dogs (*Canis familiaris*) and Cats (*Felis catus*)." *Journal of Nutrition* 136 (2006): 1927S–1931S.

Bradshaw, John, and Elizabeth Paul. "Could Empathy for Animals Have Been an Adaptation in the Evolution of *Homo sapiens*?" *Animal Welfare* 19, suppl. 1 (2010): 107–12.

Braithwaite, Victoria. *Do Fish Feel Pain?* New York: Oxford University Press, 2010.

Braitman, Laurel. *Animal Madness: How Anxious Dogs, Compulsive Parrots, and Elephants in Recovery Help Us Understand Ourselves*. New York: Simon & Schuster, 2014.

Brandow, Michael. *A Matter of Breeding: A Biting History of Pedigree Dogs and How the Quest for Status Has Harmed Man's Best Friend*. Boston: Beacon Press, 2015.

Brestrup, Craig. *Disposable Animals: Ending the Tragedy of Throwaway Pets*. Leander, TX: Camino Bay Books, 1997.

Brewer, Nathan. "The History of Euthanasia." *Lab Animal* 11 (1982): 17–19.

Broom, D. "Cognitive Ability and Awareness in Domestic Animals and Decisions about Obligations to Animals. *Applied Animal Behaviour Science* 126 (2010): 1–11.

Bryant, Clifton D. "The Zoological Connection: Animal-Related Human Behavior." *Social Forces* 58 (1979): 399.

Bukowski, J. A., and D. Wartenberg. "An Alternative Approach for Investigating the Carcinogenicity of Indoor Air Pollution: Pets as Sentinels of Environmental Cancer Risk." *Environmental Health Perspectives* 105 (1997): 1312–19.

Burgess-Jackson, Keith. "Doing Right by Our Companion Animals." *Journal of Ethics* 2 (1998): 159–85.

Burghardt, Gordon M. "Environmental Enrichment and Cognitive Complexity in Reptiles and Amphibians: Concepts, Review, and Implications for Captive Populations." *Applied Animal Behaviour Science* 147 (2013): 286–98.

Bush, Emma R., Sandra E. Baker, and David W. Macdonald. "Global Trade in Exotic Pets, 2006–2012." *Conservation Biology* 28 (2014): 663–76.

Caplan, Arthur. "Organ Transplantation." In *From Birth to Death and Bench to Clinic: The Hastings Center Bioethics Briefing Book for Journalists, Policymakers, and Campaigns*, ed. Mary Crowley, 129–32. Garrison, NY: Hastings Center, 2008.

Caras, Roger A. *A Perfect Harmony: The Intertwining Lives of Animals and Humans Throughout History*. West Lafayette, IN: Purdue University Press, 2002.

Carbone, Larry. *What Animals Want: Expertise and Advocacy in Laboratory Animals Welfare Science*. New York: Oxford University Press, 2004.

Cardinal, George. *The Rat*. New York: Wiley Publishing 2001.

Carlisle-Frank, Pamela, and Rom Flanagan. *Silent Victims: Recognizing and Stopping Abuse of the Family Pet*. Lanham, MD: University Press of America, 2006.

Carruthers, Peter. *The Animals Issue: Moral Theory in Practice*. Cambridge: Cambridge University Press, 1992.

Cassidy, Rebecca. "Zoosex and Other Relationships with Animals." In *Transgressive Sex: Subversion and Control in Erotic Encounters*, edited by Hastings Donnan and Fiona Magowan, 91–112. New York: Berghahn Books, 2009.

Cinquepalmi, Vittoria, Rosa Monno, Luciana Fumarola, Gianpiero Ventrella, Carla Calia, Maria Fiorella Greco, Danila de Vito, and Leonardo Soleo. "Environmental Contamination by Dog's Faeces: A Public Health Problem?" *International Journal of Environmental Research and Public Health*. 10 (2013): 72–84.

Cohen, Stanley. *States of Denial: Knowing about Atrocities and Suffering*. Cambridge: Polity Press, 2002.

Coleman, Phyllis. "Man['s Best Friend] Does Not Live By Bread Alone: Imposing a Duty to Provide Veterinary Care." *Animal Law* 12 (2005): 7–37.

Coren, Stanley. *How to Speak Dog: Mastering the Art of Dog-Human Communication*. New York: Free Press, 2000.

———. *The Modern Dog: A Joyful Exploration of How We Live with Dogs Today*. New York: Free Press, 2008.

Daniels, Thomas J., and Marc Bekoff. "Domestication, Exploitation, and Rights." In *Interpretation and Explanation in the Study of Animal Behavior*. Vol. 1, *Interpretation, Intentionality, and Communication*, edited by Marc Bekoff and Dale Jamieson, 345–77. Boulder, CO: Westview Press, 1990.

Davis, Susan E., and Margo DeMello. *Stories Rabbits Tell: A Natural and Cultural History of a Misunderstood Creature*. New York: Lantern Books, 2003.

DeGrazia, David. *Animal Rights: A Very Short Introduction*. Oxford: Oxford University Press, 2002.

———. "The Ethics of Confining Animals: From Farms to Zoos to Human Homes." In *The Oxford Handbook of Animal Ethics*, edited by Tom L. Beauchamp and R. G. Frey, 738–68. New York: Oxford University Press, 2011.

———. *Taking Animals Seriously: Mental Life and Moral Status*. New York: Cambridge University Press, 1996.

Dekkers, Midas. *Dearest Pet: On Bestiality*. Translated by Paul Vincent. New York: Verso, 1994.

De Laroche, Robert, and Jean-Michel Labat. *The Secret Life of Cats*. Hauppauge, NY: Barron's, 1997.

Delise, Karen. *The Pit Bull Placebo: The Media, Myths and Politics of Canine Aggression*. [United States]: Anubis Publishing, 2007.

DeLoache, J. S., M. B. Pickard, and V. LoBue. "How Very Young Children Think about Animals." In *How Animals Affect Us: Examining the Influence of Human–*

animal Interaction on Child Development and Human Health, edited by P. McCardle, S. McCune, J. A. Griffin, and V. Maholmes, 85–99. Washington, DC: American Psychological Association, 2011.

Derr, Mark. "Dog Breeds." In *Encyclopedia of Human-Animal Relations: A Global Exploration of Our Connections with Animals*, edited by Marc Bekoff, 633–39. Westport, CT: Greenwood Press, 2007.

———. *How the Dog Became the Dog*. New York: Overlook Press, 2011.

———. "The Politics of Dogs: Criticism of the Policies of AKC." *Atlantic* 265, no. 3 (1990): 49.

DeSilva, Sena S., and Giovanni M. Turchini. "Towards Understanding the Impacts of the Pet Food Industry on World Fish and Seafood Supplies." *Journal of Agricultural and Environmental Ethics* 21 (2008): 459–67.

DeTora, Michael, and Robert J. McCarthy. "Ovariohysterectomy versus Ovariectomy for Elective Sterilization of Female Dogs and Cats: Is Removal of the Uterus Necessary?" *Journal of the American Veterinary Medical Association* 239 (2011): 1409–12.

Donaghue, Emma. *Room*. Boston: Back Bay Books, 2011.

Donovan, Josephine, and Carol J. Adams, eds. *The Feminist Care Tradition in Animal Ethics: A Reader*. New York: Columbia University Press, 2007.

Driscoll, Carlos A., Juliet Clutton-Brock, Andrew C. Kitchener, and Stephen J. O'Brien. "The Taming of the Cat." *Scientific American* 300 (2009): 68–75.

Eddy, Timothy J. "What Is a Pet?" *Anthrozoös* 16 (2003): 98–122.

Edwards, Jenny. "Webinar: Understanding and Prosecuting Bestiality," ASPCA Professional, January 14, 2014, http://www.aspcapro.org/webinar/2014-01-14/understanding-and-prosecuting-bestiality.

Endenburg, Nienke, and Ben Baarda, "The Role of Pets in Enhancing Human Well-Being: Effects on Child Development." In *The Waltham Book of Human-Animal Interactions: Benefits and Responsibilities*, edited by I. Robinson, 7–18. Terrytown, NY: Elsevier Science, 1995.

Falk, Armin, and Nora Szech. "Morals and Markets." *Science* 340 (2013): 707–11.

Fiddes, Nick. *Meat: A Natural Symbol*. New York: Routledge, 1991.

Flynn, Clifton P. *Understanding Animal Abuse: A Sociological Analysis*. New York: Lantern Books, 2012.

Fogle, Bruce. *Interrelations between People and Pets*. Springfield, IL: Charles C. Thomas, 1981.

Food and Drug Administration/Center for Veterinary Medicine Report on the Risk from Pentobarbital in Dog Food. Silver Spring, MD: U.S. Food and Drug Administration, February 28, 2002. http://www.fda.gov/aboutfda/centersoffices/officeoffoods/cvm/cvmfoiaelectronicreadingroom/ucm129131.htm.

Fossat, Pascal, Julien Bacqué-Cazenave, Phillippe De Deurwaerdère, Jean-Paul

Delbecque, and Daniel Cattaert. "Anxiety-like Behavior in Crayfish Is Controlled by Serotonin." *Science* 344 (2014): 1293–97.

Fox, Michael. "Relationships between the Human and Non-human Animals." In *Interrelations between People and Pets*, edited by Bruce Fogle, 23–40. Springfield, IL: Charles C. Thomas, 1981.

———. *Returning to Eden: Animal Rights and Human Responsibility*. New York: Viking Press, 1980.

Fox, Michael W., Elizabeth Hodgkins, and Marion E. Smart. *Not Fit for a Dog! The Truth about Manufactured Dog and Cat Food*. Fresno, CA: Quill Driver Books, 2009.

Francis, Richard C. *Domesticated: Evolution in a Man-Made World*. New York: W. W. Norton & Company, 2015.

Franklin, Jon. *The Wolf in the Parlor: How the Dog Came to Share Your Brain*. New York: St. Martin's Griffin, 2009.

Fraser, David. *Understanding Animal Welfare: The Science in Its Cultural Context*. Ames, IA: Wiley-Blackwell, 2008.

Frischmann, Carol. *Pets and the Planet: A Practical Guide to Sustainable Pet Care*. Hoboken, NJ: Wiley, 2009.

Fudge, Erika. *Pets*. New York: Routledge, 2014.

Galtung, Johan. *Peace by Peaceful Means: Peace and Conflict, Development and Civilization*. Thousand Oaks, CA: Sage Publications, 1996.

Ghirlanda, S., A. Acerbi, and H. Herzog. "A Case Study in Media Influence on Choice: Dog Movie Stars and Dog Breed Popularity." *PLoS ONE* 9, no. 9 (2014): e106565. doi:10.1371/journal.pone.0106565.

Gilbey, Andrew, and Kawtar Tani. "Companion Animals and Loneliness: A Systematic Review of Quantitative Studies." *Anthrozoös* 28, no. 2 (2015): 181–97.

Gladwell, Malcolm. 2006. "Troublemakers: What Pit Bulls Can Teach Us about Profiling." *New Yorker*, February 6, 2006. http://www.newyorker.com/archive/2006/02/06/060206fa_fact.

Gompper, Matthew E., ed. *Free-Ranging Dogs and Wildlife Conservation*. New York: Oxford University Press, 2014.

Grandin, Temple, and Catherine Johnson. *Animals Make Us Human: Creating the Best Life for Animals*. Boston: Houghton Mifflin Harcourt, 2009.

Grebowicz, Margret. "When Species Meat: Confronting Bestiality Pornography." *humanimalia* 1 (2010): 1–17.

Grier, Katherine C. *Pets in America: A History*. New York: Harcourt, 2006.

Griffin, James, Sandra McCune, Valerie Maholmes, and Karyl Hurley. "Human-Animal Interaction Research: An Introduction to Issues and Topics." In *How Animals Affect Us: Examining the Influence of Human-Animal Interaction on Child Development and Human Health*, edited by Peggy McCardle, Sandra

McCune, James A. Griffin, and Valerie Maholmes, 3–10. Washington, DC: American Psychological Association, 2011.

Grimm, David. *Citizen Canine: Our Evolving Relationship with Cats and Dogs*. New York: Public Affairs Publisher, 2014.

Gruen, Lori. *The Ethics of Captivity*. New York: Oxford University Press, 2014.

Gullone, Eleonora. *Animal Cruelty, Antisocial Behavior and Aggression*. New York: Palgrave Macmillan, 2012.

———. "An Evaluative Review of Theories Related to Animal Cruelty." *Journal of Animal Ethics* 4 (2014): 37–57.

Gullone, Eleonora, and John P. Clarke. "Animal Abuse, Cruelty, and Welfare: An Australian Perspective." In *The International Handbook of Animal Abuse and Cruelty: Theory, Research, and Application*, edited by Frank R. Ascione, 305–34. West Lafayette, IN: Purdue University Press, 2008.

Hanauer, David A., Naren Ramakrishnan, and Lisa S. Seyfried. "Describing the Relationship between Cat Bites and Human Depression Using Data from an Electronic Health Record." *PLOS One* 8 (2013): e70585.

Hankin, Susan. "Making Decisions about Our Animals' Health Care: Does It Matter Whether We Are Owners or Guardians?" *Stanford Journal of Animal Law and Policy* 2 (2009): 2–51.

Haraway, Donna. *The Companion Species Manifesto: Dogs, People, and Significant Otherness*. Chicago: Prickly Paradigm Press, 2003.

———. *When Species Meet*. Minneapolis: University of Minnesota Press, 2008.

Hare, Brian, and Michael Tomasello. "Human-like Social Skills in Dogs?" *Trends in Cognitive Sciences* 9 (2005): 439–44.

Hare, Brian, and Vanessa Woods. *The Genius of Dogs: How Dogs Are Smarter Than You Think*. New York: Plume, 2013.

Harker, Rachael M., Glyn M. Collis, and June McNicholas. "The Influence of Current Relationships upon Pet Animal Acquisition." In *Companion Animals and Us: Exploring the Relationships between People and Pets*, edited by Anthony L. Podberscek, Elizabeth S. Paul, and James A. Serpell, 189–208. New York: Cambridge University Press, 2000.

Harman, Elizabeth. "The Moral Significance of Animal Pain and Animal Death." In *The Oxford Handbook of Animal Ethics*, edited by Tom L. Beauchamp and R. G. Frey, 726–37. New York: Oxford University Press, 2011.

Harriman, Marinell. *House Rabbit Handbook: How to Live with an Urban Rabbit*. Alameda, CA: Drollery Press, 1995.

Harris Interactive. "Pets Aren't Just Animals; They Are Members of the Family." September 13, 2012. http://www.harrisinteractive.com/NewsRoom/HarrisPolls/tabid/447/ctl/ReadCustom%20Default/mid/1508/ArticleId/1076/Default.aspx.

Havey, Julia, Frances R. Vlasses, Peter H. Vlasses, Patti Ludwig-Beymer, and

Diana Hackbarth. "The Effect of Animal-Assisted Therapy on Pain Medication Use after Joint Replacement." *Anthrozoös* 27 (2014): 361–69.

Hawthorne, Mark. *Bleating Hearts: The Hidden World of Animal Suffering*. Washington, DC: Change Maker Books, 2013.

Headley, Bruce, and Markus Grabka. "Health Correlates of Pet Ownership from National Surveys." In *How Animals Affect Us: Examining the Influence of Human-Animal Interaction on Child Development and Human Health*, edited by Peggy McCardle, Sandra McCune, James A. Griffin, and Valerie Maholmes, 153–62. Washington, DC: American Psychological Association, 2011.

Hensley, Christopher, Suzanne E. Tallichet, and Stephen D. Singer. "Exploring the Possible Link between Childhood and Adolescent Bestiality and Interpersonal Violence." *Journal of Interpersonal Violence* 21 (2006): 910–23.

Herzog, Hal. "Biology, Culture, and the Origins of Pet-Keeping." *Animal Behaviour and Cognition* 1 (2015): 296–308.

———. *Some We Love, Some We Hate, Some We Eat*. New York: Harper Perennial, 2010.

Hetts, Suzanne, Dan Estep, and Amy R. Marder. "Psychological Well-Being in Animals." In *Mental Health and Well-Being in Animals*, edited by Franklin D. McMillan, 211–20. Ames, IA: Blackwell Publishing Professional, 2005.

Hetts, Suzanne, Marsha L. Heinke, and Daniel Q. Estep. "Behavior Wellness Concepts for General Veterinary Practice." *Journal of the American Veterinary Medical Association* 225 (2004): 506–13.

Hoffman, Jessica M., Kate E. Creevy, and Daniel E. L. Promislow. "Reproductive Capability Is Associated with Lifespan and Cause of Death in Companion Dogs." *PLoS One* (2013). doi: 10.1371/journal.pone.0061082.

Homans, John. *What's a Dog For? The Surprising History, Science, Philosophy, and Politics of Man's Best Friend*. New York: Penguin Press, 2012.

Horowitz, Alexandra. "*Canis Familiaris*: Companion or Captive?" In *The Ethics of Captivity*, edited by Lori Gruen, 7–21. New York: Oxford University Press, 2014.

———. *Inside of a Dog: What Dogs See, Smell, and Know*. New York: Scribner, 2010.

Hribal, Jason. *Fear of the Animal Planet*. Petaluma, CA: CounterPunch, 2010.

Hua, Yaowu, Songmei Huc, Weilin Wangc, Xiaohong Wud, Fiona B. Marshalle, Xianglong Chena, Liangliang Houa, and Changsui Wanga. "Earliest Evidence for Commensal Processes of Cat Domestication." *Proceedings of the National Academy of Sciences* 11 (2014): 116–20.

Irvine, Leslie. *If You Tame Me: Understanding Our Connection with Animals*. Philadelphia: Temple University Press, 2004.

———. *My Dog Always Eats First: Homeless People and Their Animals*. Boulder, CO: Lynne Rienner Publishers, 2013.

Johnson, Rebecca A. "Animal-Assisted Intervention in Health Care Contexts." In *How Animals Affect Us: Examining the Influence of Human-Animal Interaction on Child Development and Human Health*, edited by Peggy McCardle, Sandra McCune, James A. Griffin, and Valerie Maholmes, 183–192. Washington, DC: American Psychological Association, 2011.

Julius, Henri, Andrea Beetz, Kurt Kotrschal, Dennis Turner, and Kerstin Uvnäs-Moberg. *Attachment to Pets: An Integrative View of Human-Animal Relationships with Implications for Therapeutic Practice*. Cambridge: Hogrefe Publishing, 2012.

Katcher, Aaron Honori, and Alan M. Beck, eds. *New Perspectives on Our Lives with Companion Animals*. Philadelphia: University of Pennsylvania Press, 1983.

Kawalek, Joseph C., Karyn D. Howard, Dorothy E. Farrell, Janice Derr, Carol V. Cope, Jean D. Jackson, and Michael J. Myers. "Effect of Oral Administration of Low Doses of Pentobarbital on the Induction of Cytochrome P450 Isoforms and Cytochrome P450-Mediated Reactions in Immature Beagles." *American Journal of Veterinary Research* 64 (2003): 1167–75.

Kay, William J., Susan P. Cohen, Herbert A. Nieberg, Carole E. Fudin, Ross E. Grey, Austin H. Kutscher, and Mohamed M. Osman. *Euthanasia of the Companion Animal: The Impact on Pet Owners, Veterinarians, and Society*. Philadelphia: Charles Press, 1988.

Kemmerer, Lisa, ed. *Sister Species: Women, Animals, and Social Justice*. Urbana: University of Illinois Press, 2011.

Kete, Kathleen. *The Beast in the Boudoir: Pet-keeping in Nineteenth Century Paris*. Oakland: University of California Press, 1995.

Kis, Anna, Ludwig Huber, and Anna Wilkinson. "Social Learning by Imitation in a Reptile (*Pogona vitticeps*)." *Animal Cognition* 18, no. 1 (2014): 325–31.

Krueger, Betsy, and Kirsten A. Kruger. *Secondary Pentobarbital Poisoning of Wildlife*. Washington, DC: U.S. Fish and Wildlife Service, 2002. http://www.fws.gov/southeast/news/2002/12-03SecPoisoningFactSheet.pdf.

Kustritz, Margaret V. Root. "Determining the Optimal Age for Gonadectomy of Dogs and Cats." *Journal of the American Veterinary Medical Association* 231 (2007): 1665–75.

Lakestani, Nelly N., Morag L. Donaldson, and Natalie Waran. "Interpretation of Dog Behavior by Children and Young Adults." *Anthrozoös* 27 (2014): 65–80.

Leigh, Diane, and Marilee Geyer. *One at a Time: A Week in an American Animal Shelter*. Santa Cruz, CA: No Voice Unheard Publishers, 2005.

Lesy, Michael. *The Forbidden Zone*. London: Pan Books, 1989.

Levinson, Boris. *Pets and Human Development*. Springfield, IL: Charles C. Thomas, 1972.

Leyhausen, Paul. *Cat Behavior: The Predatory and Social Behavior of Domestic and*

Wild Cats. Translated by Barbara A. Tonkin. Garland Series in Ethology. New York: Garland STPM Press, 1979.

Linzey, Andrew, ed. *The Global Guide to Animal Protection*. Urbana: University of Illinois Press, 2013.

———, ed. *The Link between Animal Abuse and Human Violence*. Portland, OR: Sussex Academies Press, 2009.

Lippmann, Walter. *Public Opinion*. New York: Free Press, 1997.

Livingston, John A. *Rogue Primate*. Boulder, CO: Roberts Rineheart Publishers, 1994.

Lorenz, Konrad. *The Waning of Humaneness*. New York: HarperCollins, 1988.

Lowe, Christopher N., Karl S. Williams, Stephen Jenkinson, and Mark Toogood. "Environmental and Social Impacts of Domestic Dog Waste in the UK: Investigating Barriers to Behavioural Change in Dog Walkers." *International Journal of Environment and Waste Management* 13 (2014): 331–47.

Mariti, Chiara, Angelo Gazzanoa, Jane Lansdown Moore, Paolo Baraglia, Laura Chellia, and Claudio Sighieria. "Perception of Dogs' Stress by Their Owners." *Journal of Veterinary Behavior: Clinical Applications and Research* 7 (2012): 213–19.

Martin, Ann N. *Food Pets Die For: Shocking Facts about Pet Food*. 3rd ed. Troutdale, OR: NewSage Press, 2008.

Maruyama, Mika and Frank R. Ascione. "Animal Abuse: An Evolving Issue in Japanese Society." In *The International Handbook of Animal Abuse and Cruelty: Theory, Research, and Application*, edited by Frank R. Ascione, 269–304. West Lafayette, IN: Purdue University Press, 2008.

Mazorlig, Tom. "Small Tanks Can Equal Big Sales." *Pet Age*, September 2014, 92–93.

McArdle, John. "Small Companion Animals: From the Lap to the Lab." *AV Magazine* (Spring 2001), 2–6.

McCardle, Peggy, Sandra McCune, James A. Griffin, and Valerie Maholmes, eds. *How Animals Affect Us: Examining the Influence of Human-Animal Interaction on Child Development and Human Health*. Washington, DC: American Psychological Association, 2011.

McConnell, Patricia B. *The Other End of the Leash: Why We Do What We Do Around Dogs*. New York: Ballantine Books, 2002.

McMahan, Jeff. *The Ethics of Killing: Problems at the Margins of Life*. New York: Oxford University Press, 2002.

McMillan, Franklin D. "Emotional Maltreatment in Animals." In *Mental Health and Well-Being in Animals*, edited by Franklin D. McMillan, 167–80. Ames, IA: Blackwell Publishing Professional, 2005.

———. "Stress, Distress, and Emotion: Distinctions and Implications for Mental

Well-Being." In *Mental Health and Well-Being in Animals*, edited by Franklin D. McMillan, 93–112. Ames, IA: Blackwell Publishing Professional, 2005.

———. "Stress-Induced Emotional Eating in Animals: A Review of the Experimental Evidence and Implications for Companion Animal Obesity." *Journal of Veterinary Behavior* 8 (2013): 376–85.

Meints, K., A. Racca, and N. Hickey. "How to Prevent Dog Bite Injuries? Children Misinterpret Dogs Facial Expressions." *Injury Prevention* 16 (2010): A68. doi:10.1136/ip.2010.029215.246.

Mellor, David, Emily Patterson-Kane, and Kevin J. Stafford. *The Sciences of Animal Welfare*. Hoboken, NJ: John Wiley & Sons, 2009.

Melson, Gail F. "Principles for Human-Animal Interaction Research." In *How Animals Affect Us: Examining the Influence of Human-Animal Interaction on Child Development and Human Health*, edited by Peggy McCardle, Sandra McCune, James A. Griffin, and Valerie Maholmes, 13–34. Washington, DC: American Psychological Association, 2011.

———. *Why the Wild Things Are: Animals in the Lives of Children*. Cambridge, MA: Harvard University Press, 2001.

Merck, Melinda D., and Doris M. Miller. "Sexual Abuse." In *Veterinary Forensics: Animal Cruelty Investigations*, edited by Melinda Merck, 233–42. 2nd ed. Oxford: Wiley-Blackwell, 2012.

Merz-Perez, Linda, and Kathleen M. Heide. *Animal Cruelty: Pathway to Violence against People*. Walnut Creek, CA: AltaMira Press, 2004.

Milani, Myrna. "Canine Surgical Sterilization and the Human-Animal Bond." In *Encyclopedia of Human-Animal Relations: A Global Exploration of Our Connections with Animals*, edited by Marc Bekoff, 919–25. Westport, CT: Greenwood Press, 2007.

Miletski, Hani. "Is Zoophilia a Sexual Orientation? A Study." In *Bestiality and Zoophilia: Sexual Relations with Animals*, edited by Andrea M. Beetz and Anthony L. Podberscek, 82–97. West Lafayette, IN: Purdue University Press, 2005.

Mills, Daniel, Emile van der Zee, and Helen Zulch. "When the Bond Goes Wrong: Problem Behaviours in the Social Context." In *The Social Dog: Behavior and Cognition*, edited by Juliane Kaminski and Sarah Marshall-Pescini, 223–45. Norwell, MA: Kluwer, 2014.

Mueller, Megan K. "Is Human-Animal Interaction (HAI) Linked to Positive Youth Development? Initial Answers." *Applied Developmental Science* 18 (2014): 5–16.

Mueller-Paul Julia, Anna Wilkinson, Ulrike Aust, Michael Steurera, Geoffrey Hall, and Ludwig Huber. "Touchscreen Performance and Knowledge Transfer in the Red-Footed Tortoise (*Chelonoidis carbonaria*)." *Behavioural Processes* 106 (2014): 187–92.

Munro, H. M. C., and M. V. Thrusfield. "'Battered Pets': Sexual Abuse." In *Bestiality and Zoophilia: Sexual Relations with Animals*, edited by Andrea M. Beetz and Anthony L. Podberscek, 71–81. West Lafayette, IN: Purdue University Press, 2005.

Nagasawa, Miho, Shouhei Mitsui, Shiori En, Nobuyo Ohtani, Mitsuaki Ohta, Yasuo Sakuma, Tatsushi Onaka, Kasutaka Mogi, and Takefumi Kikusui. "Oxytocin-Gaze Positive Loop and the Coevolution of Human-Dog Bonds." *Science* 348 (2015): 333–36.

Natterson-Horowitz, Barbara, and Kathryn Bowers. *Zoobiquity: What Animals Can Teach Us about Health and the Science of Healing*. New York: Alfred A. Knopf, 2012.

Nibert, David A. *Animal Oppression and Human Violence: Domesecration, Capitalism, and Global Conflict*. New York: Columbia University Press, 2013.

———. *Animal Rights/Human Rights: Entanglements of Oppression and Liberation*. Lanham, MD: Rowman & Littlefield, 2002.

Nicoll, Kate. *Soul Friends: Finding Healing with Animals*. Indianapolis, IN: Dog Ear Publishing, 2005.

Norrgran, Jessica, Bernt Jones, Anders Bignert, Ioannis Athanassiadis, and Åke Bergman. "Higher PBDE Serum Concentrations May Be Associated with Feline Hyperthyroidism in Swedish Cats." *Environmental Science and Technology* 49 (2015): 5107–14.

Noske, Barbara. *Beyond Boundaries: Humans and Animals*. Buffalo, NY: Black Rose Books, 1997.

Oldfield, Ronald G. "Aggression and Welfare in a Common Aquarium Fish, the Midas Cichlid." *Journal of Applied Animal Welfare Science* 14 (2011): 340–60.

Overall, Christine. *Ethics and Human Reproduction: A Feminist Analysis*. Boston: Allen and Unwin, 1987.

Overall, Karen L. *Clinical Behavioral Medicine for Small Animals*. Saint Louis: Mosby, 1997.

———. "Mental Illness in Animals—the Need for Precision in Terminology and Diagnostic Criteria." In *Mental Health and Well-Being in Animals*, edited by Franklin D. McMillan, 127–44. Ames, IA: Blackwell Publishing Professional, 2005.

Pagani, Camilla, Francesco Robustelli, and Frank R. Ascione. "Animal Abuse Experiences Described by Italian School-Aged Children." In *The International Handbook of Animal Abuse and Cruelty: Theory, Research, and Application*, edited by Frank R. Ascione, 247–68. West Lafayette, IN: Purdue University Press, 2008.

Palmer, Clare. *Animal Ethics in Context*. New York: Columbia University Press, 2010.

Palmer, Clare, Sandra Corr, and Peter Sandoe. "Inconvenient Desires: Should We Routinely Neuter Companion Animals?" *Anthrozoös* 25 (2012): s153–s172.

Palmer, Clare, and Peter Sandøe. "For Their Own Good: Captive Cats and Routine Confinement." In *The Ethics of Captivity*, edited by Lori Gruen, 135–55. New York: Oxford University Press, 2014.

Panksepp, Jaak. *Affective Neuroscience: The Foundations of Human and Animal Emotions*. New York: Oxford University Press, 2004.

Panksepp, Jaak, and Lucy Bevin. *The Archaeology of Mind: Neuroevolutionary Origins of Human Emotions*. New York: W. W. Norton, 2012.

Patronek, Gary. "Animal Hoarding: A Third Dimension of Animal Abuse." In *The International Handbook of Animal Abuse and Cruelty: Theory, Research, and Application*, edited by Frank R. Ascione, 221–46. West Lafayette, IN: Purdue University Press, 2008.

Paul, Elizabeth S. "Empathy with Animals and with Humans: Are They Linked? *Anthrozoös* 13 (2000): 194–202.

Pedersen, Helena. *Animals in Schools: Processes and Strategies in Human-Animal Education*. West Lafayette, IN: Purdue University Press, 2010.

Peeters, Marijke E., and Jolle Kirpensteijn. "Comparison of Surgical Variables and Short-Term Postoperative Complications in Healthy Dogs Undergoing Ovariohysterectomy or Ovariectomy." *Journal of the American Veterinary Medical Association* 238 (2011): 189–94.

Phillips, Allie. *Defending the Defenseless: A Guide to Protecting and Advocating for Pets*. Lanham, MD: Rowman & Littlefield, 2011.

———. *How Shelter Pets Are Brokered for Experimentation*. Lanham, MD: Rowman & Littlefield, 2010.

Pierce, Jessica. *The Last Walk: Reflecting on Our Pets at the End of Their Lives*. Chicago: University of Chicago Press, 2012.

Podberscek, Anthony L., Elizabeth S. Paul, and James A. Serpell. *Companion Animals and Us: Exploring the Relationships between People and Pets*. New York: Cambridge University Press, 2000.

Pongrácz, Péter, Csaba Molnár, Antal Dóka, and Ádám Miklósi. "Do Children Understand Man's Best Friend? Classification of Dog Barks by Pre-Adolescents and Adults." *Applied Animal Behaviour Science* 135 (2011): 95–102. doi: http://dx.doi.org/10.1016/j.applanim.2011.09.005.

Poresky, R. H. "The Young Children's Empathy Measure: Reliability, Validity and Effects of Companion Animal Bonding." *Psychological Reports* 66 (1990): 931–36.

Previde, Emanuela Prato, and Paola Valsecchi. "The Immaterial Cord: The Dog-Human Attachment Bond." In *The Social Dog: Behavior and Cognition*, edited by Juliane Kaminski and Sarah Marshall-Pescini, 165–89. Norwell, MA: Kluwer, 2014.

Ramos, D., A. Reche-Junior, P. L. Fragoso, R. Palme, N. K. Yanasse, V. R. Gouvêa, A. Beck, and D. S. Mills. "Are Cats (*Felis catus*) from Multi-Cat Households More Stressed? Evidence from Assessment of Fecal Glucocorticoid Metabolite Analysis." *Physiology and Behavior* 122 (2013): 72–75.

Range, Friederike, Caroline Ritter, and Zsófia Virányi. "Testing the Myth: Tolerant Dogs and Aggressive Wolves." *Proceedings of the Royal Society B* 282 (2015). doi: 10.1098/rspb.2015.0220.

Range, Friederike, and Zsófia Virányi, "Tracking the Evolutionary Origins of Dog-Human Cooperation: The 'Canine Cooperation Hypothesis.'" *Frontiers in Psychology* (2015). http://journal.frontiersin.org/article/10.3389/fpsyg.2014.01582/full.

Regan, Tom. *The Case for Animal Rights*. Oakland: University of California Press, 1983.

Robin, Michael, and Robert ten Bensel. "Pets and the Socialization of Children." *Marriage and Family Review* 8 (1985): 63–78.

Rollin, Bernard E. *Animal Rights and Human Morality*. 3rd ed. Amherst, NY: Prometheus Books, 2006.

Rothman, Barbara K. *Recreating Motherhood: Ideology and Technology in a Patriarchal Society*. New York: Norton, 1989.

Salvanes A. G. V., O. Moberg, L. O. E. Ebbesson, T. O. Nilsen, K. H. Jensen, and V. A. Braithwaite. "Environmental Enrichment Promotes Neural Plasticity and Cognitive Ability in Fish." *Proceedings of the Royal Society B: Biological Sciences* (2013) 280 (1767): 20131331. doi: 10.1098/rspb.2013.1331.

Sapontzis, Steven. *Morals, Reasons, and Animals*. Philadelphia: Temple University Press, 1987.

Savishinsky, Joel S. "Common Fate, Difficult Decision: A Comparison of Euthanasia in People and in Animals." In *Euthanasia of the Companion Animal: The Impact on Pet Owners, Veterinarians, and Society*, edited by William J. Kay et al., 3–8. Philadelphia: Charles Press, 1988.

Schoen, Allen M. *Kindred Spirits: How the Remarkable Bond between Humans and Animals Can Change the Way We Live*. New York: Broadway Books, 2001.

Seibert, Lynne. "Mental Health Issues in Captive Birds." In *Mental Health and Well-Being in Animals*, edited by Franklin D. McMillan, 285–94. Ames, IA: Blackwell Publishing Professional, 2005.

Selhub, Eva M., and Alan C. Logan. *Your Brain on Nature: The Science of Nature's Influence on Your Health, Happiness, and Vitality*. Mississauga, Ontario: John Wiley & Sons, 2012.

Serpell, James A. "Creatures of the Unconscious: Companion Animals as Mediators." In *Companion Animals and Us: Exploring the Relationships between People and Pets*, edited by Anthony L. Podberscek, Elizabeth S. Paul, and James A. Serpell, 108–21. New York: Cambridge University Press, 2000.

———. "Humans, Animals, and the Limits of Friendship." In *The Dialectics of Friendship*, edited by Roy Porter and Sylvana Tomaselli, 111–29. New York: Routledge, 1989.

———. *In the Company of Animals: A Study of Human-Animal Relationships*. New York: Cambridge University Press, 1996.

———. "Pet-Keeping and Animal Domestication: A Reappraisal." In *The Walking Larder: Patterns of Domestication, Pastoralism and Predation*, edited by J. Clutton-Brock, 10–21. London: Unwin Hyman, 1989.

Serpell, James, and Elizabeth Paul. "Pets and the Development of Positive Attitudes to Animals." In *Animals and Human Society: Changing Perspectives*, edited by Aubrey Manning and James Serpell, 127–44. New York: Routledge, 1994.

———. "Pets in the Family: An Evolutionary Perspective." In *The Oxford Handbook of Evolutionary Family Psychology*, edited by Catherine A. Salmon and Todd K. Shackelford, 297–309. New York: Oxford University Press, 2011.

Shapiro, Kenneth Joel. *Animal Models of Human Psychology: Critique of Science, Ethics, and Policy*. Seattle, WA: Hogrefe & Huber Publishers, 1998.

Shapiro, Michael H. "Fragmenting and Reassembling the World: Of Flying Squirrels, Augmented Persons, and Other Monsters." *Ohio State Law Journal* 51 (1990): 331–74.

Sheldrake, Rupert. *Dogs That Know When Their Owners Are Coming Home*. Rev. ed. New York: Three Rivers Press, 2011.

Shir-Vertesh, Dafna. "'Flexible Personhood': Loving Animals as Family Members in Israel." *American Anthropologist* 114 (2012): 420–32.

Sinclair, Leslie, Melinda Merck, and Randall Lockwood. *Forensic Investigation of Animal Cruelty: A Guide for Veterinary and Law Enforcement Professionals*. Washington, DC: Humane Society Press, 2006.

Singer, Peter. "Heavy Petting." *Nerve* (2011). http://www.utilitarianism.net/singer/by/2001——.htm.

Sonntag, Q., and K. L. Overall. "Key Determinants of Dog and Cat Welfare: Behaviour, Breeding, and Household Lifestyle." *Scientific and Technical Review of the Office International des Epizooties* (Paris) 33, no. 1 (2014): 213–20.

Speigel, Marjorie. *The Dreaded Comparison: Human and Animal Slavery*. New York, New York: Mirror Books, 1996.

Špinka, Marek, and Françoise Wemelsfelder. "Environmental Challenge and Animal Agency." In *Animal Welfare*, edited by Michael C. Appleby, Joy A. Mench, I. Anna S. Olsson, and Barry O. Hughes, 27–43. 2nd ed. Cambridge, MA: CAB, 2011.

Steiner, Gary. *Animals and the Moral Community: Mental Life, Moral Status, and Kinship*. New York: Columbia University Press, 2008.

Stull, Jason, Jason Brophy, and J. S. Weese. "Reducing the Risk of Pet-Associated Zoonotic Infections." *Canadian Medical Association Journal* (2015). doi: 10.1503/cmaj.141020.

Szasz, Kathleen. *Petishism: Pet Cults of the Western World*. London: Hutchinson & Co., 1968.

Takaoka, Akiko, Tomomi Maeda, Yusuke Hori, and Kazuo Fujit. "Do Dogs Follow Behavioral Cues from an Unreliable Human?" *Animal Cognition* 18 (2015): 475–83.

Téglás, E., A. Gergely, K. Kupán, A. Miklósi, and J. Topál. "Dog's Gaze Following Is Tuned to Human Communicative Signals." *Current Biology* 22 (2012): 209–12.

Tegzes, John, Oscar E. Chavez, Broc A. Sandelin, and Lee Allen Pettey. "Just Food for Dogs White Paper: An Evidence-Based Analysis of the Dog Food Industry in the USA." JustFoodForDogs LLC, Newport Beach, CA, 2014. http://truthaboutpetfood.com/JFFDwp.pdf.

Thixton, Susan. *Buyer Beware: The Crimes, Lies and Truth about Pet Food*. Lexington, KY: [CreateSpace], 2011.

Thomas, Keith. *Man and the Natural World: Changing Attitudes in England, 1500–1800*. New York: Oxford University Press, 1996.

Thurston, Mary Elizabeth. *The Lost History of the Canine Race: Our 15,000-Year Love Affair with Dogs*. Riverside, NJ: Andrews McMeel Publishers, 1996.

Torres, Bob. *Making a Killing: The Political Economy of Animal Rights*. Oakland: AK Press, 2007.

Torres de la Riva G., B. L. Hart, T. B. Farver, A. M. Oberbauer, L. L. M. Messam, et al. "Neutering Dogs: Effects on Joint Disorders and Cancers in Golden Retrievers." *PLoS ONE* 8 (2013): e55937. doi:10.1371/journal.pone.0055937 2013.

Tuan, Yi-Fu. *Dominance and Affection: The Making of Pets*. New Haven, CT: Yale University Press, 1984.

Twiest, Mark G., Meghan Mahoney Tweist, and Mary Rench Jalongo. "A Friend at School: Classroom Pets and Companion Animals in the Curriculum." In *The World's Children and Their Companion Animals: Developmental and Educational Significance of the Child/Pet Bond*, edited by Mary Renck Jalongo, 61–78. Olney, MD: Association for Childhood Education International.

Varner, Gary. "Pets, Companion Animals, and Domesticated Partners." In *Ethics for Everyday*, edited by David Benatar, 450–75. New York: McGraw-Hill, 2002.

Vermeulen, H., and J. S. J. Odendaal. "Proposed Typology of Companion Animal Abuse." *Anthrozoös* 6 (1993): 248–57.

Vittoria Cinquepalmi, Rosa Monno, Luciana Fumarola, Gianpiero Ventrella, Carla Calia, Maria Fiorella Greco, Danila de Vito, and Leonardo Soleo. "Environmental Contamination by Dog's Faeces: A Public Health Problem?" *International Journal of Environmental Research and Public Health* 10 (2013): 72–84.

Vrontou, Sophia, Allan M. Wong, Kristofer K. Rau, H. Richard Koerber, and David J. Anderson. "Genetic Identification of C Fibres That Detect Massage-Like Stroking of Hairy Skin *in vivo*." *Nature* 493 (2013): 669–73.

Wakefield, Jerome. "DSM-5 Proposed Criteria For Sexual Paraphilias: Tensions between Diagnostic Validity and Forensic Utility." *International Journal of Law and Psychiatry* 34 (2011): 195–209.

Warwick, Clifford. "The Morality of the Reptile 'Pet' Trade." *Journal of Animal Ethics* 4 (2014): 74–94.

Wathan, Jennifer, and Karen McComb. "The Eyes and Ears Are Visual Indicators of Attention in Domestic Horses." *Current Biology* 24 (2014): R677–R679.

Weese, J. Scott, and Martha B. Fulford. *Companion Animal Zoonoses*. Ames, IA: Wiley-Blackwell, 2011.

Weil, Zoe. "Humane Education." In *Encyclopedia of Human-Animal Relations: A Global Exploration of Our Connections with Animals*, edited by Marc Bekoff, 675–78. Westport, CT: Greenwood Press, 2007.

Welfare in Dog Training. "What's Wrong with Using 'Dominance' to Explain the Behaviour of Dogs?" (2013). Available at: http://www.dogwelfarecampaign.org/why-not-dominance.php.

Wemelsfelder, Françiose. "Animal Boredom: Understanding the Tedium of Confined Lives." In *Mental Health and Well-Being in Animals*, edited by Franklin D. McMillan, 79–92. Ames, IA: Blackwell Publishing Professional, 2005.

Wilkins, Adam S., Richard W. Wrangham, and W. Tecumseh Fitch. "The 'Domestication Syndrome' in Mammals: A Unified Explanation Based on Neural Crest Cell Behavior and Genetics." *Genetics* 197 (2014): 795–808.

Williams, Marta. *Ask Your Animal: Resolving Behavioral Issues through Intuitive Communication*. Novato, CA: New World Library, 2008.

Wilson, Cindy C., and Dennis C. Turner, eds. *Companion Animals in Human Health*. Thousand Oaks, CA: Sage Publications, 1998.

Wright, Phyllis. "Why Must We Euthanize?" 1978; repr., *Shelter Sense* 18, no. 9 (October 1995): 7–8.

Wrye, Jen. "Beyond Pets: Exploring Relational Perspectives of Petness." *Canadian Journal of Sociology* 34 (2009): 1033–63.

Wynne, Clive D. L., Nicole R. Dorey, and Monique A. R. Udell. "The Other Side of the Bond: Domestic Dogs' Human-Like Behavior." In *How Animals Affect Us: Examining the Influence of Human-Animal Interaction on Child Development and Human Health*, edited by Peggy McCardle, Sandra McCune, James A. Griffin, and Valerie Maholmes, 101–16. Washington, DC: American Psychological Association, 2011.

Wynne Tyson, Jon. *The Extended Circle: A Commonplace Book of Animal Rights*. New York: Paragon House, 1989.

Yin, Sophia. *How to Behave So Your Dog Behaves*. Neptune City, NJ: T. F. H. Publications, 2004.

———. *The Perfect Puppy in 7 Days: How to Start Your Puppy Off Right*. Davis, CA: Cattle Dog Publishing, 2011.

Young, Robert. J. *Environmental Enrichment for Captive Animals*. Oxford: Blackwell Science, 2003.

Index